Prepare for State [barcode sticker] with ArgoPrep

Award Winning Program

1st Grade Math Practice Book

8 Full-Length Simulated Common Core Math Tests

ArgoPrep is one of the leading providers of supplemental educational products and services. We offer affordable and effective test prep solutions to educators, parents and students. Learning should be fun and easy! To access more resources visit us at www.argoprep.com.

Our goal is to make your life easier, so let us know how we can help you by e-mailing us at: info@argoprep.com.

ISBN: 9781951048259
Published by ArgoPrep.

- ArgoPrep is a recipient of the prestigious **Mom's Choice Award**.
- ArgoPrep also received the 2019 **Seal of Approval** from Homeschool.com for our award-winning workbooks.
- ArgoPrep was awarded the 2019 **National Parenting Products Award**, **Gold Medal Parent's Choice Award** and **the Tillywig Brain Child Award.**

BOOKS BY ARGOPREP

Here are some other test prep workbooks by ArgoPrep you may be interested in. All of our workbooks come equipped with detailed video explanations to make your learning experience a breeze! Visit us at www.argoprep.com

COMMON CORE MATH SERIES

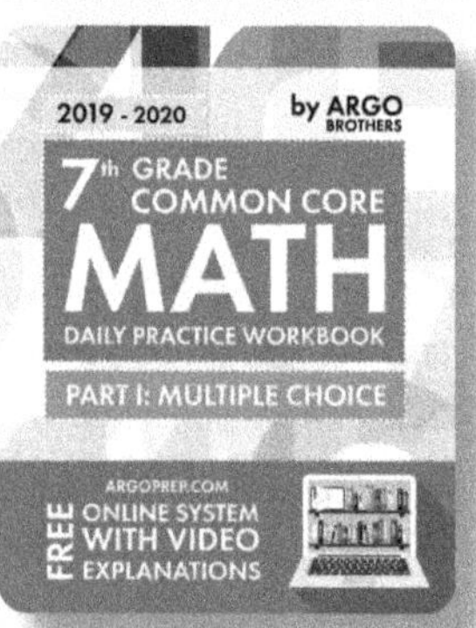

COMMON CORE ELA SERIES

COMMON CORE TEST SERIES

The goal of these workbooks is to provide mock state tests so students can increase confidence and test scores during actual test day.

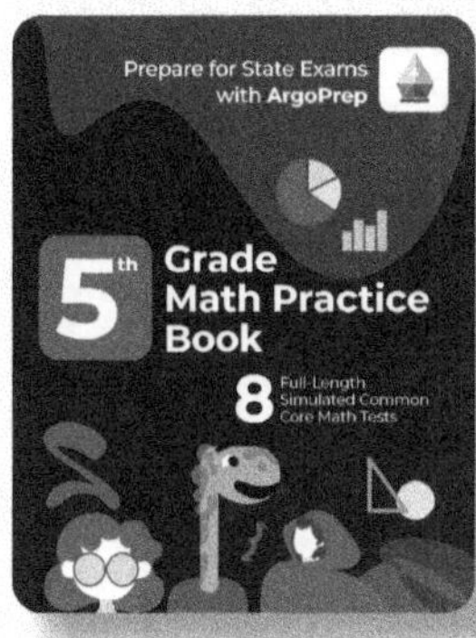

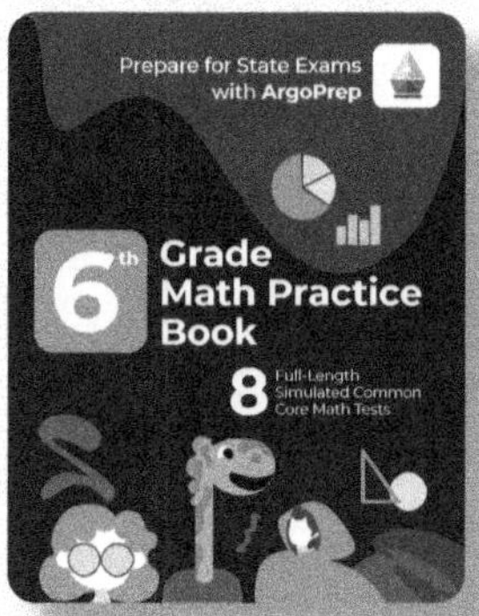

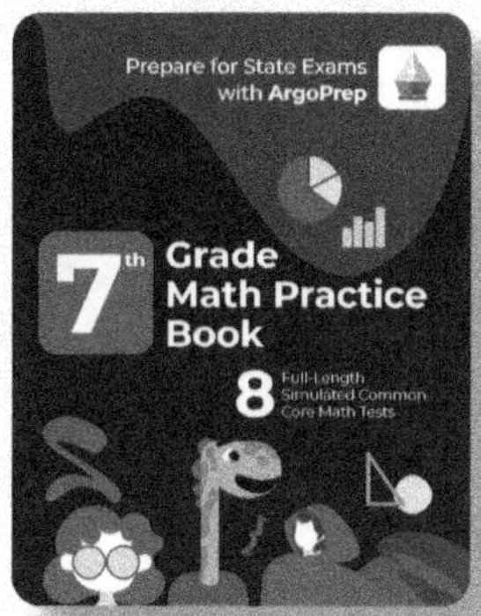

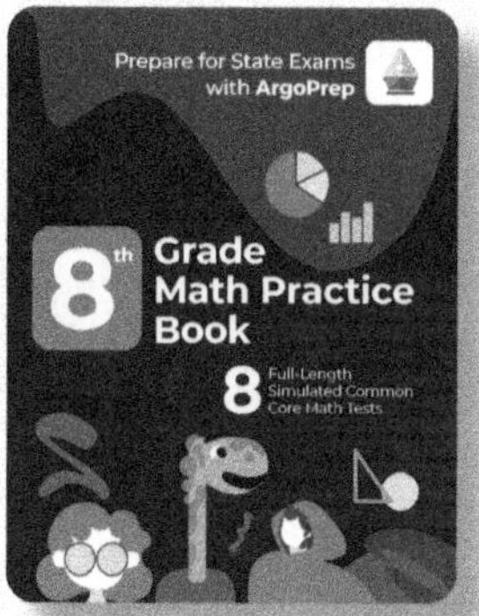

INTRODUCING MATH!

Introducing Math! by ArgoPrep is an award-winning series created by certified teachers to provide students with high-quality practice problems. Our workbooks include topic overviews with instruction, practice questions, answer explanations along with digital access to video explanations. Practice in confidence - with ArgoPrep!

KIDS SUMMER ACADEMY SERIES

ArgoPrep's Kids Summer Academy series helps prevent summer learning loss and gets students ready for their new school year by reinforcing core foundations in math, english and science. Our workbooks also introduce new concepts so students can get a head start and be on top of their game for the new school year!

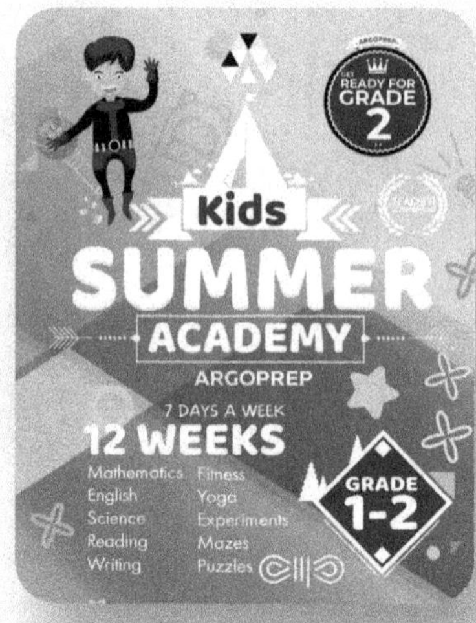

SUMMER ACTIVITY PLAYGROUND SERIES

Summer Activity Playground is another summer series that is designed to prevent summer learning loss and prepares students for the new school year. Students will be able to practice math, ELA, science, social studies and more! This is a new released series that offers the latest aligned learning standards for each grade.

Take a look inside

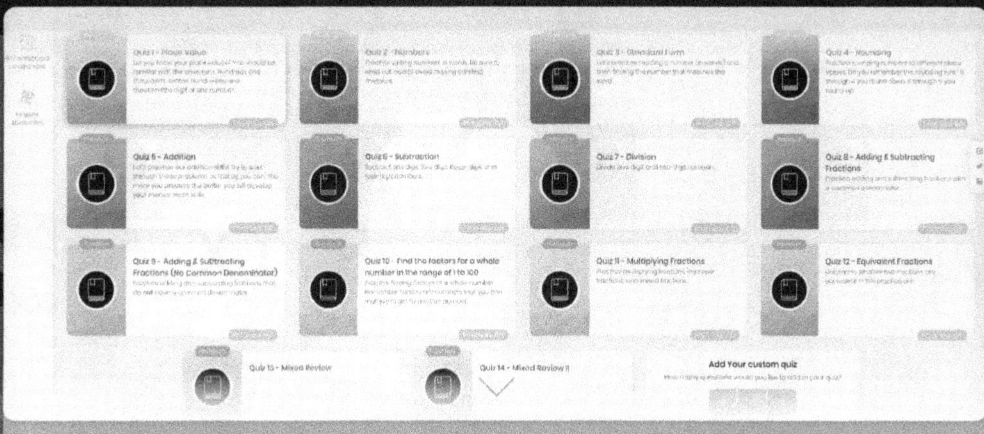

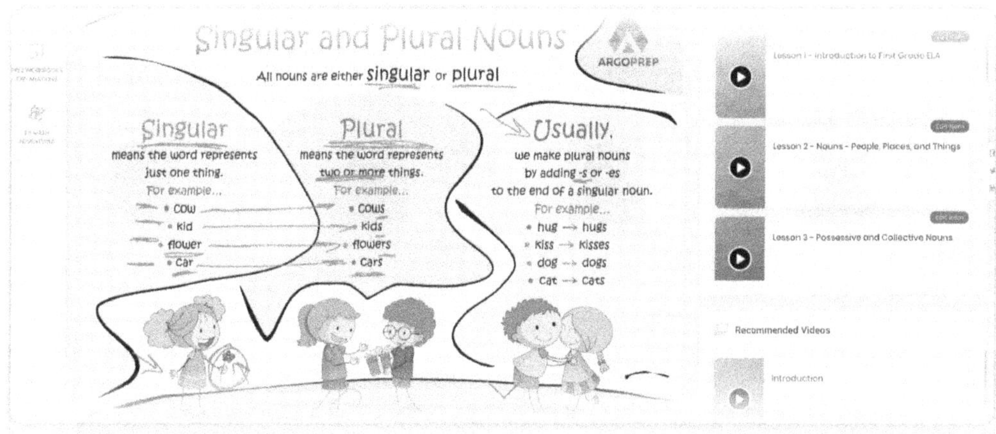

Testimonials

Read all testimonials

I love these books, both the multiple choice and the free response. They are quite challenging math problems that really require critical thinking, and a true understanding of the relationship between numbers. I am a homeschooling parent, and have used these books for 3rd grade and now 4th grade math. These books provide excellent practice especially in word problems across all topic areas.

Samanta K.

1 month ago

Awesome practice to brainstorm your kids' math knowledge; keeps my son's attention and interest to work with this book!!! He loves it! Definitely a good buy!

Olga Kuznec

2 month ago

Your annual subscription includes **two free** workbooks shipped **directly to your home**.

Kindergarten 1st grade 2nd grade 3rd grade 4th grade 5th grade 6th grade 7th grade 8th grade

Subscribe for 12 months and get 2 workbooks **for free**

$ 119 88

save 40%

SUBSCRIBE

Child Safe Parent Monitoring

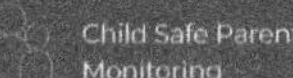

No Ads

No External Links

Table of contents

How to use the book

This workbook contains eight full-length practice exams that simulate state exams for **Grade 1**. The goal of this workbook is to provide mock state tests so students can increase confidence and test scores during actual test day. The exams mirror New York State Exams and reflect the latest learning standards tested for the **2020-2021** school year. If you are a teacher using this workbook, you can find the common core standards for each question in the answer key.

ArgoPrep recommends users of this workbook to simulate real testing conditions to provide the maximum benefit. Students should aim to complete an entire session in a sitting just like in the state exam. Please provide students with extra scrap paper while working through the exams.

For a detailed overview of the Common Core State Standards for 1st grade, please visit: www.corestandards.org/Math/Content/1/introduction/

Have any questions? Contact us at: **info@argoprep.com**

Test 1: Session 1

Tips: Be sure to read each question carefully and work through the problem step by step before choosing the correct answer.

Suggested Time:

Session 1: 55 to 65 Minutes
Session 2: 60 to 75 minutes

Students are **NOT** permitted to use a calculator.

1. There are 4 cows in the barn and 1 more joins them. How many cows are there all together?

A. 6
B. 4
C. 5
D. 3

2. Add 2 + 5 = ________

A. 6
B. 7
C. 8
D. 4

3. How do you make 7?

A. 10 - 3
B. 10 - 4
C. 9 - 4
D. 9 - 3

4. If 4 + 5 = 9, then 9 - 5 = ________

A. 4
B. 5
C. 9
D. 18

5. Miss Baker has 3 daisies, 4 sunflowers and 2 roses. How many flowers does she have all together?

A. 10
B. 8
C. 9
D. 7

6. Add the double numbers. What does 3 + 3 = ?

A. 9
B. 5
C. 4
D. 6

GO ON ▶

7. How do you make 8?

A. 3 + 4
B. 5 + 3
C. 5 + 2
D. 3 + 6

8. Subtract. 7 - 0 = ?

A. 0
B. 8
C. 7
D. 6

9. Fill in the missing number. 5 + _______ = 10

A. 4
B. 3
C. 7
D. 5

10. Complete the double fact 8 = ________ + ________

A. 3, 3
B. 2, 2
C. 4, 4
D. 5, 5

11. Fill in the missing number. 9 - 4 = ________ ?

A. 5
B. 4
C. 12
D. 13

12. Add. 2 + 3 = ________

A. 4
B. 5
C. 6
D. 7

GO ON ▶

13. What is the missing fact family?

6 + 2 = 8	2 + 6 = 8	8 - 2 = 6

A. 2 + 6 = 8
B. 2 + 4 = 8
C. 8 - 6 = 2
D. 8 - 4 = 4

14. How do you write **FIVE** using digits?

A. 6
B. 7
C. 8
D. 5

15. How many tens are in 22?

A. 2
B. 3
C. 22
D. 4

16. How many ones are in 37?

A. 3
B. 7
C. 6
D. 10

17. Add. 40 + 11 = ________ ?

A. 50
B. 61
C. 51
D. 41

18. Subtract. 30 - 20 = ________ ?

A. 50
B. 20
C. 30
D. 10

GO ON

19. What is ten more than 7?

A. 27
B. 17
C. 10
D. 7

20. What is ten less than 34?

A. 14
B. 44
C. 24
D. 4

21. Add. 8 + 8 + 2 = ________ ?

A. 18
B. 19
C. 17
D. 20

22. Add. 17 + 4 = ________ ?

A. 20
B. 19
C. 18
D. 21

23. 34 is 3 tens and 4 ones. What is another way to make 34?

A. 2 tens and 3 ones
B. 4 tens and 3 ones
C. 3 tens and 14 ones
D. 2 tens and 14 ones

24. What time does the clock show?

A. 12 : 00
B. 6 : 00
C. 7 : 00
D. 1 : 00

GO ON

25. Joanie bought fruit at the store. How many oranges did she buy?

FRUIT					
Oranges	 🍊	 🍊	 🍊	 🍊	 🍊
Pineapples	 🍍	 🍍			

A. 7 **B.** 2 **C.** 5 **D.** 6

Session 2

GO ON 

26. Which eraser is the longest? Circle the eraser that is the longest

A.
B.
C.

27. At recess Bobby took a poll to see who likes soccer or baseball better. His results are in the table. How many people liked soccer best?

Best Sport		
How many?	14	11

A. 10 B. 11 C. 13 D. 14

28. Which shape is the cube?

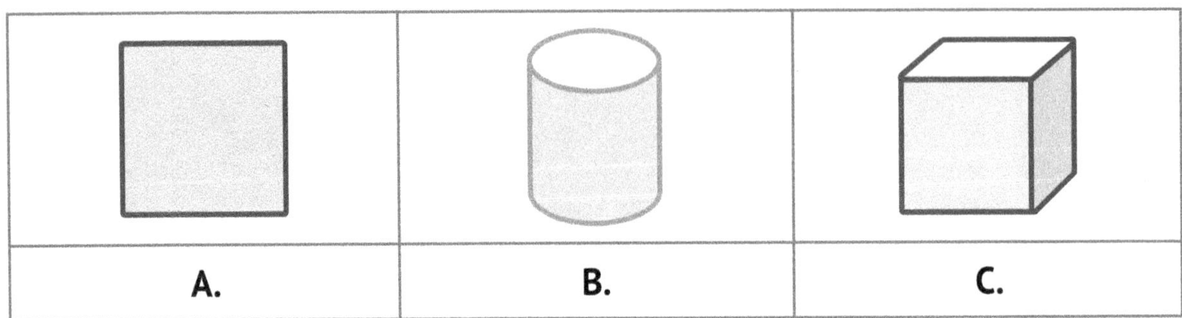

A.	B.	C.

29. Which shape has three sides?

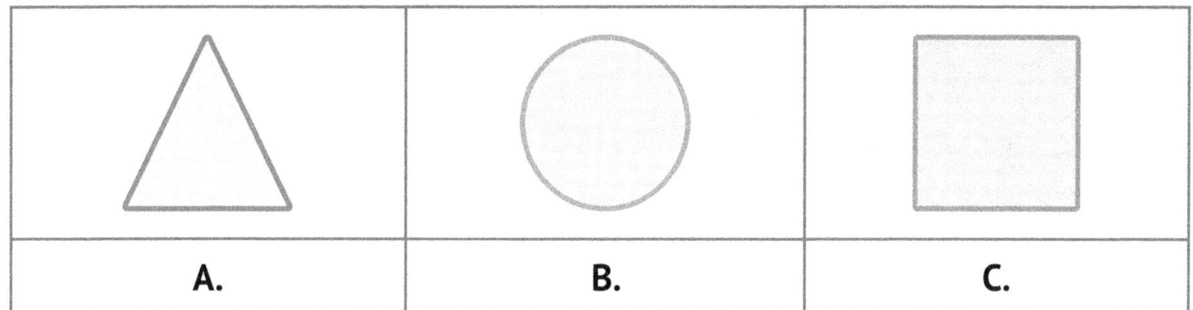

30. There are 6 cars in the parking lot at school. If one more drives in, how many cars are there in total?

A. 1
B. 6
C. 7
D. 5

31. Jeffrey has 3 green pencils, 5 red pencils and 2 blue pencils. How many pencils does he have all together?

A. 10
B. 9
C. 6
D. 11

GO ON ▶

32. What is the missing fact family?

3 + 2 = 5	5 - 2 = 3	2 + 3 = 5

A. 5 + 3 = 8
B. 5 - 3 = 2
C. 5 - 4 = 1
D. 5 + 2 = 7

33. What time on the clock matches the times below?

A.	B.	C.	D.
6 : 00	6 : 30	3 : 30	4 : 30

34. Use >, < or = to make the sentence true.

34 ______ 44

35. 26 is 2 tens and six ones. What is another way to make 26?

____

____

36. How much money is in the box? ________

37. Fill in the missing number. If 3 + 6 = 9, then 9 - 6 = ________

GO ON

38. Johnny has 8 cookies and he gives 5 to Betsy. How many cookies does he have left?

____

39. Abby has 3 dogs and Jessie has 4 dogs. How many dogs do they have together? Write an addition sentence that fits this story.

____

40. Kayla used her allowance to buy candy canes. She spent $5 and has $2 left. How much money did she start with?

____

Test 2: Session 1

Tips: Be sure to read each question carefully and work through the problem step by step before choosing the correct answer.

Suggested Time:

Session 1: 55 to 65 Minutes
Session 2: 60 to 75 minutes

Students are **NOT** permitted to use a calculator.

1. Add. 4 + 4 = ________

A. 4
B. 6
C. 8
D. 10

2. There were 9 dogs at the shelter, 9 dogs were adopted. How many are still at the shelter?

A. 1
B. 0
C. 9
D. 5

3. Counting forward starting at 72, what number comes next?

A. 70
B. 71
C. 73
D. 74

4. How many ones are in 16?

A. 1
B. 3
C. 7
D. 6

5. How many stars are in the frame?

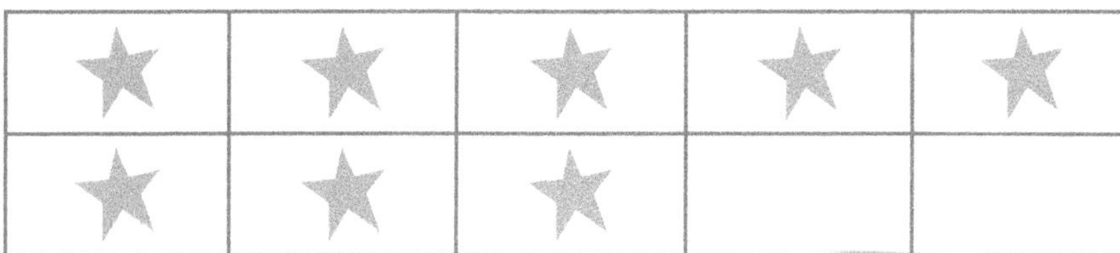

A. 10 **B.** 9 **C.** 7 **D.** 8

6. If 4 + 5 = 9, then ______

A. 9 - 5 = 4
B. 9 + 5 = 4
C. 5 + 9 = 4
D. 4 + 9 = 5

GO ON ▶

7. What is the number TEN using digits?

A. 4
B. 6
C. 8
D. 10

8. Subtract. 90 - 10 = ________

A. 90
B. 80
C. 70
D. 60

9. Add. 5 + 2 + 5 = ________

A. 10
B. 15
C. 12
D. 13

10. There are 3 turtles in a pond, 2 more turtles crawl in, how many turtles are there all together?

A. 6
B. 8
C. 5
D. 1

11. Match the analog clock to the digital time.

A. 5 : 00
B. 6 : 00
C. 12 : 00
D. 12 : 30

12. Subtract. 13 - 13 = ______

A. 13
B. 0
C. 1
D. 4

GO ON

13. Joe has 14 lemons, he gives Sarah 7 lemons, how many lemons does Joe have left?

A. 3
B. 6
C. 7
D. 9

14. What subtraction sentence is true?

A. 10 - 4 = 5
B. 10 - 6 = 4
C. 10 - 5 = 4
D. 10 - 7 = 6

15. How do you make 3?

A. 10 -7 =
B. 10 - 6 =
C. 9 - 3 =
D. 10 - 5 =

16. What coin is this?

A. Nickel
B. Dime
C. Penny
D. Quarter

17. Add. 4 + 2 + 1 = ________

A. 10
B. 5
C. 6
D. 7

18. Add. 20 + 40 = ________

A. 40
B. 30
C. 50
D. 60

GO ON ▶

19. What is the doubles fact? 10 = _____ + _____

A. 5, 5
B. 6, 6
C. 4, 4
D. 3, 3

20. How do you write 17 in words?

A. Sixteen
B. Seventeen
C. Seventy
D. Sixty

21. How many tens are in 41?

A. 1
B. 2
C. 3
D. 4

22. How many ones are in 41?

A. 1
B. 2
C. 3
D. 4

23. How do you make 7?

A. 3 + 4 =
B. 3 + 3 =
C. 3 + 2 =
D. 5 + 4 =

24. What addition sentence is true?

A. 4 + 7 = 11
B. 4 + 7 = 12
C. 4 + 8 = 11
D. 4 + 6 = 11

GO ON

25. How much money is in the box?

A. $.12

B. $.15

C. $.11

D. $.13

Session 2

GO ON

26. Add. 12 + 4 = ______

A. 12
B. 14
C. 16
D. 18

27. Find the missing number 10 - ______ = 4

A. 4
B. 5
C. 6
D. 7

28. Add. 7 + 8 = ______

A. 15
B. 14
C. 13
D. 16

29. Fill in the blank. If 3 + 6 = 9, Then 9 - _____ = 3

A. 3
B. 6
C. 9
D. 18

30. Add. 6 + 2 + 3 = _____?

A. 9
B. 10
C. 12
D. 11

31. Irene went to the grocery store and bought 3 bananas, 2 oranges and 5 apples, how much fruit did she buy in all?

A. 8
B. 9
C. 10
D. 11

GO ON

32. Subtract. 17 - 17 = ______?

A. 0
B. 1
C. 17
D. 34

33. What is the missing number 2 + ______ = 7 ?

A. 4
B. 5
C. 6
D. 7

34. What symbol makes this sentence true 34 ______ 34?

35. What shape is this?

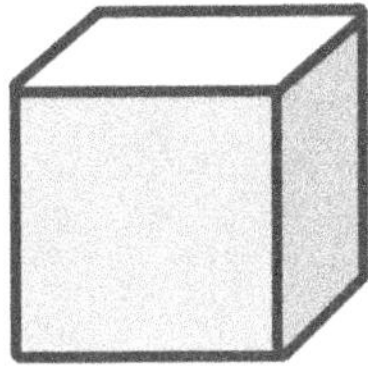

____

36. What time does the clock read?

____

37. Write the missing fact family.

1 + 3 = 4	4 - 1 = 3	4 - 3 = 1

____

GO ON ▶

38. What is 10 more than 26?

____

39. Gwen has 2 hamsters and 3 guinea pigs. How many pets does she have in all?

____

40. Jack has 17 teddy bears. He gives Lulu 9 teddy bears. How many teddy bears does Jack have left?

Write a subtraction sentence that fits this story.

____

Test 3: Session 1

Tips: Be sure to read each question carefully and work through the problem step by step before choosing the correct answer.

Suggested Time:

Session 1: 55 to 65 Minutes
Session 2: 60 to 75 minutes

Students are **NOT** permitted to use a calculator.

1. What is the number 7 using words?

A. Seven
B. Six
C. Four
D. Three

2. How many hearts are in the frame?

♥	♥	♥	♥	♥
♥				

A. 6 **B.** 5 **C.** 4 **D.** 10

3. Add. 5 + 5 = ______

A. 0
B. 5
C. 8
D. 10

4. How many tens are in 12?

A. 0
B. 2
C. 1
D. 3

5. There are 4 hippos doing a hula dance, 2 more join in, how many hippos are dancing together?

A. 6
B. 5
C. 2
D. 1

6. Add. 30 + 50 = ______

A. 60
B. 80
C. 100
D. 40

GO ON

7. Subtract. 12 - 0 = ______

A. 12
B. 0
C. 6
D. 4

8. Add. 6 + 1 + 6 = ______

A. 6
B. 8
C. 12
D. 13

9. What is 10 more than 54?

A. 54
B. 44
C. 74
D. 64

10. Add. 3 + 3 + 1 = ______

A. 7
B. 8
C. 9
D. 10

11. How many tens are in 56?

A. 7
B. 6
C. 9
D. 5

12. How many ones are in 56?

A. 7
B. 6
C. 9
D. 5

GO ON

13. What is the number fourteen using digits?

A. 14
B. 40
C. 45
D. 44

14. What subtraction sentence is true?

A. 14 - 7 = 6
B. 14 - 4 = 0
C. 14 - 4 = 10
D. 14 - 3 = 10

15. What coin is this?

A. Nickel
B. Dime
C. Penny
D. Quarter

16. Mary has 13 crackers, she eats 5 of them, how many crackers does she have left?

A. 8
B. 7
C. 18
D. 17

17. What addition sentence is true?

A. 6 + 5 = 11
B. 6 + 5 = 12
C. 6 + 6 = 11
D. 6 + 4 = 11

18. What is the missing number 7 - ______ = 1?

A. 7
B. 6
C. 5
D. 4

GO ON

19. How do you make 5?

A. 9 - 3 =
B. 9 - 2 =
C. 10 - 4 =
D. 9 - 4 =

20. Counting forward starting at 49, what number comes next?

A. 48
B. 50
C. 51
D. 52

21. How do you make 9?

A. 2 + 6 =
B. 5 + 4 =
C. 2 + 5 =
D. 5 + 6 =

22. Jen has 17 dolls, she gave Kate 8 dolls, how many dolls does Jen have left?

A. 7
B. 9
C. 11
D. 13

23. What would make this sentence true 33 _____ 43?

A. >
B. <
C. =

24. What is the double fact that makes the sentence true 8 = _____ + _____

A. 8, 8
B. 3, 3
C. 4, 4
D. 5, 5

GO ON ▶

25. What time does the clock read?

A. 12 : 30

B. 2 : 30

C. 3 : 30

D. 6 : 30

Session 2

GO ON 

26. Add 11 + 9 = ______

A. 10
B. 20
C. 19
D. 18

27. Fill in the blank. If 8 - 3 = 5, the 5 + ______ = 8

A. 3
B. 5
C. 8
D. 0

28. Add. 2 + 3 + 4 = ______

A. 6
B. 7
C. 8
D. 9

29. How do you make 10?

A. 3 + 6 = 10
B. 9 + 1 = 10
C. 9 + 2 = 10
D. 3 + 8 = 10

30. Subtract. 18 - 9 = ______?

A. 9
B. 10
C. 11
D. 12

31. What is the missing number 2 + ______ = 9?

A. 4
B. 5
C. 6
D. 7

GO ON

32. Add. 11 + 6 = ______

A. 10
B. 13
C. 15
D. 17

33. Bear the dog has 3 stuffed toys, 4 tennis balls and 3 ropes. How many toys does he have all together?

A. 8
B. 9
C. 10
D. 11

34. What shape has 3 corners?

____

35. How much money is in the box? ______________

36. What time does the clock read?

____

37. Subtract 80 - 20 = ______

GO ON

38. Write the missing fact family.

5 - 3 = 2	5 - 2 = 3	3 + 2 = 5

____

39. Fill in the missing number. If 10 + 4 = 14, then 14 - _____ = 10

40. Tommy has 16 rubber duckies; he leaves 10 in the bathtub and takes the rest to bed. How many rubber duckies does he take to bed?

Write a subtraction sentence for this story.

____

Test 4: Session 1

Tips: Be sure to read each question carefully and work through the problem step by step before choosing the correct answer.

Suggested Time:

Session 1: 55 to 65 Minutes
Session 2: 60 to 75 minutes

Students are **NOT** permitted to use a calculator.

1. How many ones are in the number 19?

A. 1
B. 10
C. 9
D. 8

2. What is ten more than 41?

A. 51
B. 61
C. 31
D. 11

3. What is the number NINE using digits?

A. 7
B. 8
C. 9
D. 10

4. Add. 1 + 3 + 5 = ______

A. 8
B. 9
C. 10
D. 11

5. What subtraction sentence is true?

A. 9 - 3 = 5
B. 9 - 4 = 4
C. 9 - 3 = 6
D. 9 - 7 = 3

6. Subtract 40 - 10 = ______

A. 50
B. 40
C. 30
D. 20

GO ON

7. Bob has 7 sunflowers, he gives 4 to Rebecca, how many sunflowers does he have left?

A. 2
B. 3
C. 4
D. 11

8. How do you write 12 in words?

A. Eleven
B. Ten
C. Twelve
D. Twenty

9. What time does the clock read?

A. 12 : 30
B. 1 : 30
C. 12 : 00
D. 5 : 00

10. Add. 2 + 2 + 7 = ______

A. 11
B. 10
C. 9
D. 8

11. Add. 10 + 40 = ______

A. 30
B. 40
C. 50
D. 60

12. How many x's are in the frames?

x	x	x	x	x
x	x	x	x	

A. 5 **B.** 8 **C.** 9 **D.** 10

GO ON

13. Add. 7 + 7 = ______

A. 14
B. 13
C. 7
D. 0

14. There are 3 moose roller and 3 more join them. How many moose are there all together?

A. 3
B. 4
C. 5
D. 6

15. How do you make 4?

A. 10 - 5 =
B. 10 - 4 =
C. 10 - 6 =
D. 9 - 6 =

16. Match the analog clock with the digital time.

A. 1 : 30
B. 6 : 30
C. 4 : 00
D. 4 : 30

17. Counting forward starting at 35, what number comes next?

A. 40
B. 34
C. 36
D. 37

18. Subtract. 14 - 1 = ______

A. 14
B. 13
C. 12
D. 11

GO ON

19. What double fact makes the sentence true _____ + _____ = 6

A. 2, 2
B. 3, 3
C. 4, 4
D. 5, 5

20. What subtraction sentence is true?

A. 5 - 5 = 0
B. 5 - 5 = 1
C. 12 - 1 = 10
D. 14 - 14 = 1

21. What is the missing number 1 + _____ = 10

A. 10
B. 9
C. 8
D. 7

22. Add. 3 + 11 = ______

A. 11
B. 12
C. 13
D. 14

23. Fill in the missing number. If 5 + 2 = 7, then 7 - ______ = 2?

A. 5
B. 3
C. 2
D. 7

24. What coin is this?

A. Nickel
B. Dime
C. Penny
D. Quarter

GO ON ▶

25. Louise has 19 colored pencils, she brings 9 colored pencils to school, how may did she leave at home?

A. 0

B. 7

C. 9

D. 10

Session 2

GO ON 

26. How do you make 5?

A. 3 + 3 =
B. 3 + 4 =
C. 2 + 3 =
D. 2 + 4 =

27. Pete the Cat has 3 tuna, 4 salmon and 1 haddock. How many fish does he have all together?

A. 8
B. 7
C. 6
D. 5

28. Counting up from 78, what number comes next?

A. 79
B. 76
C. 80
D. 88

29. Add. 14 + 5 = ______

A. 15
B. 19
C. 17
D. 13

30. Add. 4 + 2 + 1 = ______

A. 5
B. 6
C. 7
D. 8

31. Subtract. 20 - 10 = ______

A. 20
B. 30
C. 5
D. 10

GO ON ▶

32. How many tens are in 29?

A. 2
B. 3
C. 8
D. 9

33. How many ones are in 29?

A. 2
B. 3
C. 8
D. 9

34. How much money is in the box? ____________________

35. Fill in the missing number. If 7 + 3 = 10, then 10 - 3 = ______

36. What is the missing number 7 - ______ = 2?

37. How many corners does this shape have? ________________

GO ON

38. Write the missing fact family.

3 + 4 = 7	4 + 3 = 7	7 - 3 = 4

____

39. Lonnie has 13 red candies and 4 pink candies. How many candies does she have all together?

____

40. What symbol would make this sentence true 55 ______ 53?

Test 5: Session 1

Tips: Be sure to read each question carefully and work through the problem step by step before choosing the correct answer.

Suggested Time:

Session 1: 55 to 65 Minutes
Session 2: 60 to 75 minutes

Students are **NOT** permitted to use a calculator.

1. Add the double fact. 4 + 4 = ______

A. 6
B. 8
C. 10
D. 12

2. Subtract. 13 - 4 = ______

A. 10
B. 9
C. 8
D. 7

3. What symbol makes this sentence true 19 ______ 17?

A. >
B. <
C. =

4. Fill in the blank. ______ - 2 = 8

A. 4
B. 6
C. 8
D. 10

5. What number completes the double fact 14 = ______ + ______?

A. 5, 5
B. 6, 6
C. 7, 7
D. 8, 8

6. What number sentence makes 7?

A. 4 + 3
B. 4 + 4
C. 3 + 5
D. 5 + 1

GO ON ▶

7. If 3 + 5 = 8, then

A. 8 - 8 = 1
B. 8 - 4 = 5
C. 8 - 5 = 3
D. 8 + 3 = 5

8. Add. 4 + 9 = ______

A. 11
B. 12
C. 13
D. 14

9. What is the related subtraction fact for 10 - 6 = 4?

A. 10 - 5 = 5
B. 10 - 4 = 6
C. 6 - 4 = 2
D. 4 - 4 = 0

10. What is eight written as a number?

A. 5
B. 6
C. 7
D. 8

11. Counting forward, what number comes after 40?

A. 45
B. 39
C. 41
D. 42

12. Add. 2 + 2 + 4 = ______

A. 6
B. 8
C. 10
D. 2

GO ON

13. Which addition sentence is true?

A. 3 + 3 = 8
B. 3 + 4 = 8
C. 4 + 3 = 7
D. 4 + 5 = 7

14. How many tens are in 46?

A. 4
B. 5
C. 6
D. 7

15. Fill in the blank. ______ + 2 = 9

A. 7
B. 5
C. 3
D. 1

16. What number is 10 more than 57?

A. 47
B. 67
C. 77
D. 87

17. What subtraction sentence is true?

A. 12 - 1 = 11
B. 12 - 2 = 11
C. 12 - 5 = 6
D. 12 - 6 = 5

18. How many ones are in 15?

A. 1
B. 3
C. 5
D. 7

GO ON

19. Add. 20 + 15 = ______

A. 30
B. 45
C. 25
D. 35

20. What time is it?

A. 12 : 30
B. 1 : 30
C. 2 : 30
D. 3 : 30

21. Fill in the blank. 8 - ______ = 0

A. 4
B. 6
C. 8
D. 0

22. Donnie went to the farm. How many cows did he see?

A. 5 **B.** 4 **C.** 3 **D.** 2

23. Subtract 8 - 3 = ______

A. 2
B. 3
C. 4
D. 5

24. What number sentence makes 9?

A. 9 - 1
B. 9 - 2
C. 9 - 0
D. 9 - 3

GO ON ▶

25. How many vertices does this shape have?

A. 2

B. 3

C. 4

D. 5

Session 2

GO ON

26. Subtract the double fact. 14 - 7 = ______

A. 14
B. 8
C. 7
D. 10

27. What is 10 less than 64?

A. 74
B. 84
C. 54
D. 44

28. Brad has 3 gummy bears and 6 Swedish fish. How much candy does he have all together?

A. 6
B. 7
C. 8
D. 9

29. What shape is this?

A. Rectangle
B. Triangle
C. Cube
D. Hexagon

30. Add. 10 + 2 = ______

A. 12
B. 11
C. 10
D. 9

31. How many tens are in this diagram?

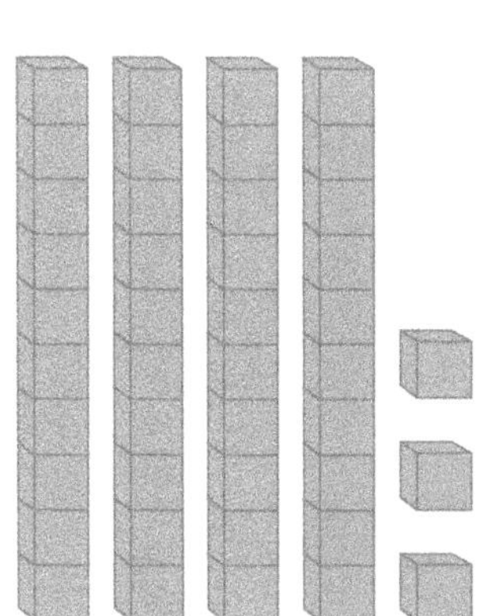

A. 3
B. 4
C. 5
D. 6

GO ON

32. Which screw is the longest? ___________________

A.

B.

C.

33. What is the related addition fact 2 + 8 = 10?

A. 10 + 0 = 10
B. 9 + 1 = 10
C. 8 + 2 = 10
D. 7 + 3 = 10

34. What number comes after 49? _________________________________

35. Tony has 15 Lego blocks, he uses 5 in his building, how many does he have left?

____

36. Fill in the blank. 7 + ______ = 11

37. What is 5 written as a word?

____

GO ON ▶

38. In the diagram, how many groups of tens are blue? ____________

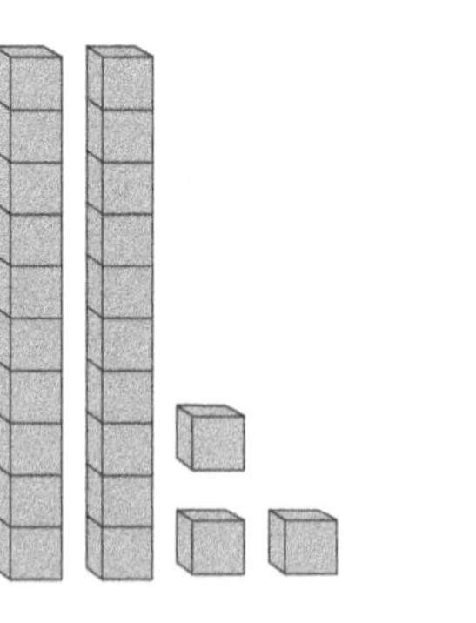

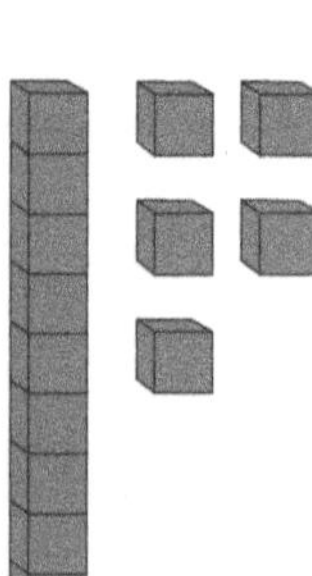

39. Jane has 7 stuffed pigs and 2 stuffed horses. How many stuffed animals does she have all together?

____

40. What is the missing part of the fact family?

1 + 4 = 5	4 + 1 = 5	5 - 4 = 1

____

Test 6: Session 1

Tips: Be sure to read each question carefully and work through the problem step by step before choosing the correct answer.

Suggested Time:

Session 1: 55 to 65 Minutes
Session 2: 60 to 75 minutes

Students are **NOT** permitted to use a calculator.

1. Add. 10 + 9 = ______

A. 17
B. 18
C. 19
D. 20

2. Fill in the blank. If 3 + 6 = 9, then 9 - ______ = 6

A. 3
B. 6
C. 9
D. 12

3. What number sentence makes 8?

A. 2 + 4
B. 2 + 6
C. 2 + 5
D. 2 + 3

4. Fill in the blank. _____ + 5 = 10

A. 3
B. 6
C. 4
D. 5

5. What is the missing fact in the fact family 2 + 7 = 9, 9 - 7 = 2, 7 + 2 = 9?

A. 9 - 2 = 7
B. 9 - 0 = 9
C. 7 + 3 = 10
D. 7 - 2 = 5

6. What is the related addition fact 3 + 4 = 7?

A. 3 + 3 = 6
B. 3 + 5 = 8
C. 4 + 4 = 8
D. 4 + 3 = 7

GO ON

7. Subtract. 7 - 2 = ____

A. 3
B. 5
C. 7
D. 9

8. Subtract the double fact. 12 - 6 = ____

A. 10
B. 8
C. 4
D. 6

9. What symbol makes this sentence true 12 ____ 14

A. >
B. <
C. =

10. Fill in the blank. 12 + ______ = 20

A. 8
B. 10
C. 18
D. 20

11. Counting forward, what number comes after 55?

A. 54
B. 53
C. 56
D. 57

12. How many ones are in 19?

A. 1
B. 3
C. 7
D. 9

GO ON ▶

13. What addition sentence is true?

A. 13 + 4 = 16
B. 13 + 4 = 17
C. 13 + 5 = 17
D. 13 + 5 = 16

14. What is ten less than 31?

A. 11
B. 21
C. 41
D. 51

15. Drew has 12 goldfish in his lunch. He shares 4 with Ben. How many does Drew have left?

A. 4
B. 6
C. 8
D. 10

16. Add. 30 + 17 = ______

A. 47
B. 27
C. 45
D. 43

17. Fill in the blank. 6 - ______ = 2

A. 2
B. 4
C. 6
D. 8

18. What time is it?

A. 5 : 30
B. 6 : 30
C. 7 : 30
D. 8 : 30

GO ON

19. What is Eleven written as a digit?

A. 10
B. 11
C. 12
D. 13

20. Joel has 8 toy dinosaurs and 9 toy frogs. How many toys does he have all together?

A. 14
B. 15
C. 16
D. 17

21. Ava takes a poll for what is the best fruit. How many people chose raspberries?

A. 17
B. 15
C. 13
D. 11

Fruit		
Number of Votes	13	17

22. What shape is this basketball?

A. Cone
B. Cube
C. Sphere
D. Triangle

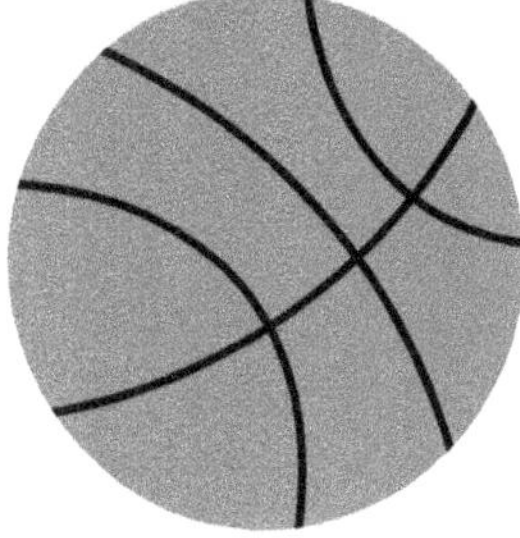

23. Add the double fact. 1 + 1 = ______

A. 1
B. 2
C. 3
D. 4

24. What subtraction sentence is true?

A. 17 - 0 = 0
B. 17 - 1 = 17
C. 17 - 2 = 15
D. 17 - 15 = 3

GO ON

25. Add. 6 + 7 = ______

A. 12

B. 13

C. 14

D. 15

Session 2

GO ON

26. What number sentence makes 6?

A. 10 - 3
B. 10 - 2
C. 10 - 1
D. 10 - 4

27. What number is 10 more than 60?

A. 50
B. 70
C. 40
D. 80

28. Add. 1 + 0 + 1 = ______

A. 10
B. 8
C. 4
D. 2

29. Which cantaloupe is the longest?

A. 

B. 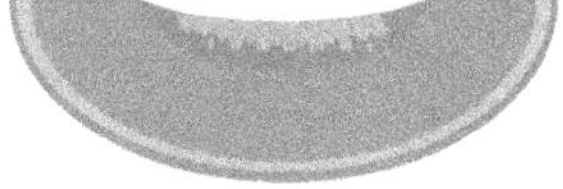

C.

30. What is the related subtraction fact 7 - 2 = 5?

A. 7 - 5 = 2
B. 5 - 2 = 3
C. 3 - 2 = 1
D. 7 - 4 = 3

31. Add. 5 + 1 = ______

A. 4
B. 5
C. 6
D. 7

GO ON ▶

32. What number completes the double fact 18 = _____ + _____ ?

A. 6, 6
B. 7, 7
C. 8, 8
D. 9, 9

33. How many vertices does this shape have?

A. 4
B. 5
C. 6
D. 7

34. What is 6 written as a word? ______________________________

35. Fill in the blank. ______ - 12 = 0

36. What number comes after 87? ______________________________

37. How many tens are 73? ______________________________

GO ON ▶

38. How many ones are in the diagram? ________________

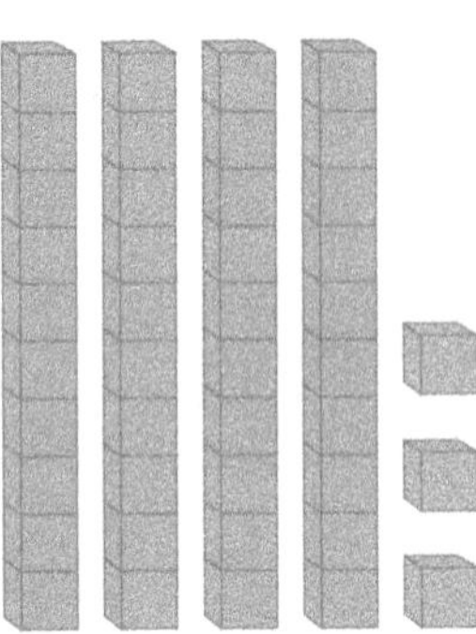

39. How many tens are in the diagram? ________________

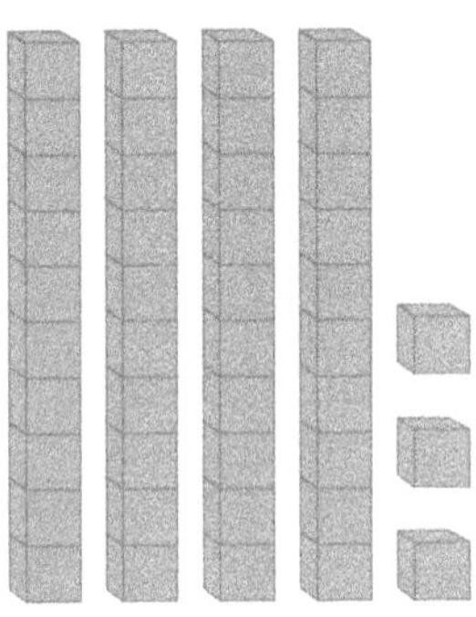

40. Bella has 9 guitars. She shares 7 with Cora. How many guitars does she have left?

____

Test 7: Session 1

Tips:

Be sure to read each question carefully and work through the problem step by step before choosing the correct answer.

Suggested Time:

Session 1: 55 to 65 Minutes
Session 2: 60 to 75 minutes

Students are **NOT** permitted to use a calculator.

1. Subtract. 6 - 3 = ______

A. 6
B. 3
C. 5
D. 2

2. What symbol makes this sentence true 14 ______ 15?

A. >
B. <
C. =

3. What number completes the double fact 10 = ______ + ______ ?

A. 4, 4
B. 5, 5
C. 6, 6
D. 7, 7

4. What is nine as a digit?

A. 8
B. 9
C. 7
D. 6

5. What addition sentence is true?

A. 18 + 1 = 20
B. 20 + 1 = 19
C. 18 + 2 = 20
D. 20 + 4 = 22

6. What subtraction sentence is true?

A. 16 - 2 = 13
B. 16 - 8 = 7
C. 16 - 9 = 10
D. 16 - 10 = 6

GO ON ▶

7. Fill in the blank. ______ - 7 = 10

A. 15
B. 17
C. 19
D. 11

8. How do you make 5?

A. 1 + 3
B. 1 + 4
C. 1 + 2
D. 2 + 1

9. Add. 4 + 1 + 3 = ______

A. 2
B. 6
C. 8
D. 10

10. Add the double fact. 8 + 8 = ______

A. 20
B. 18
C. 16
D. 14

11. What is the missing fact in the fact family 1 + 9 = 10, 10 - 9 = 1, 10 - 1 = 9?

A. 10 + 0 = 10
B. 9 - 1 = 8
C. 10 - 0 = 10
D. 9 + 1 = 10

12. Subtract. 10 - 2 = ______

A. 6
B. 8
C. 10
D. 12

13. Iris saw 7 fish swimming in the pond, but 1 swam away. How many fish were left?

A. 6
B. 5
C. 4
D. 3

14. Counting forward, what number comes after 99?

A. 98
B. 87
C. 97
D. 100

15. How many tens are in 61?

A. 1
B. 2
C. 4
D. 6

16. Add. 40 + 21 = ______

A. 41
B. 51
C. 61
D. 71

17. What time does the clock read?

A. 1 : 30
B. 2 : 30
C. 12 : 30
D. 6 : 30

18. Fill in the blank. 8 - ______ = 7

A. 0
B. 1
C. 2
D. 3

GO ON ▶

19. Liz has 4 red shoes and 7 pink shoes. How many shoes does she have all together?

A. 10
B. 11
C. 12
D. 13

20. How do you make 3?

A. 9 - 3
B. 9 - 4
C. 9 - 5
D. 9 - 6

21. Joe went to a barbeque. Below is how many hot dogs and hamburgers he ate. How many hot dogs did Joe eat?

A. 2
B. 3
C. 4
D. 5

22. Subtract. 17 - 0 = ______

A. 0
B. 5
C. 15
D. 17

23. What is the related addition fact 5 + 1 = 6?

A. 1 + 6 = 7
B. 1 + 5 = 6
C. 5 + 6 = 11
D. 5 + 0 = 5

24. What number is 10 less than 73?

A. 43
B. 53
C. 63
D. 83

GO ON ▶

25. Fill in the blank. 6 + ______ = 6

A. 6
B. 4
C. 2
D. 0

Session 2

GO ON

26. Add. 9 + 3 = ______

A. 9
B. 10
C. 11
D. 12

27. How many edges does a triangle have?

A. 5
B. 4
C. 3
D. 2

28. How many tens are in the diagram?

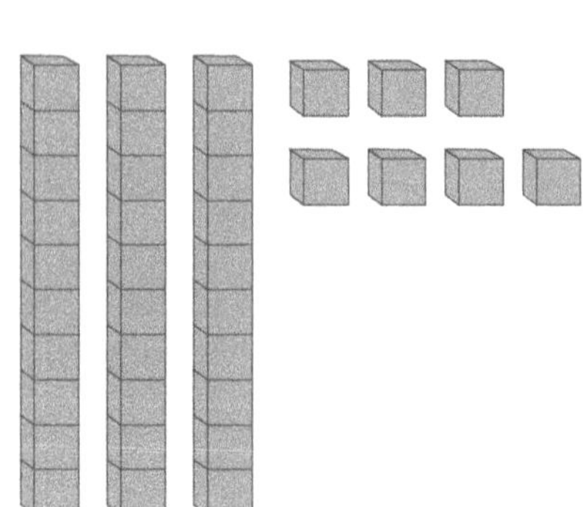

A. 3
B. 5
C. 6
D. 7

29. How many ones are in the diagram?

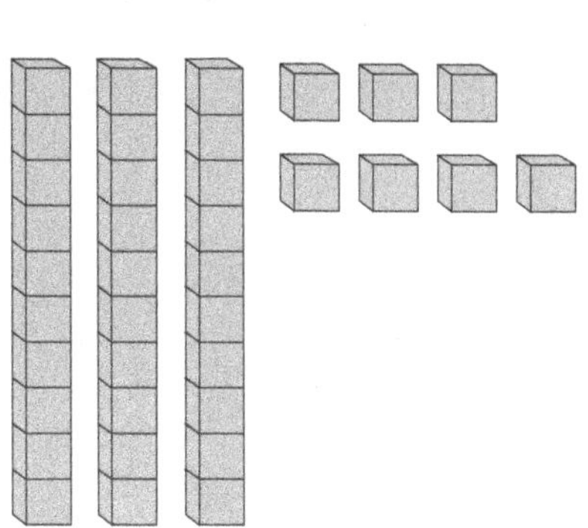

A. 3
B. 5
C. 6
D. 7

30. How many edges does a sphere have?

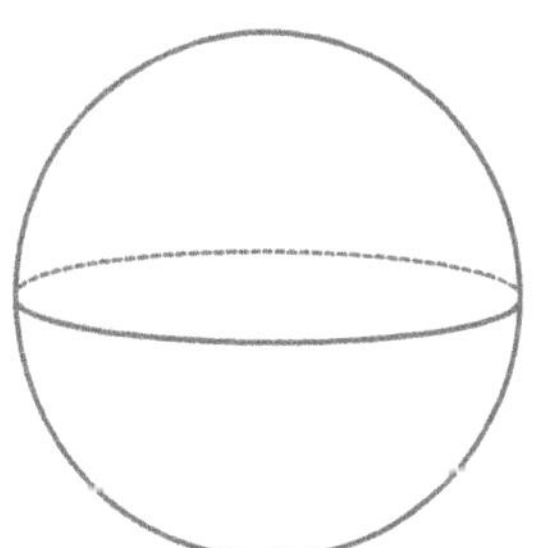

A. 0
B. 1
C. 2
D. 3

31. Mike took a poll to see if his classmates liked dogs or cats better. How many students liked cats better?

A. 8
B. 9
C. 11
D. 12

Animals		
Number of Votes	12	9

GO ON ▶

32. What is 10 more than 84?

A. 64
B. 74
C. 54
D. 94

33. Subtract the double fact. 20 - 10 = ____

A. 10
B. 20
C. 30
D. 40

34. Fill in the missing number. If 10 - 3 = 7, then 3 + ____ = 10

35. What number comes after 29? ______

36. What is the missing number 12 + ______ = 14?

37. Write 10 as a word ______

GO ON

38. What is the related subtraction fact 6 - 2 = 4?

____

39. How many ones are in 13?

____

40. Matt has 9 coins. He threw 4 of them in the wishing well. How many coins does he have left?

____

Test 8: Session 1

Tips: Be sure to read each question carefully and work through the problem step by step before choosing the correct answer.

Suggested Time:

Session 1: 55 to 65 Minutes
Session 2: 60 to 75 minutes

Students are **NOT** permitted to use a calculator.

1. What number sentence makes 4?

A. 8 - 7
B. 8 - 3
C. 8 - 5
D. 8 - 4

2. Teddy saw 12 crows in the field, 5 flew away. How many are left?

A. 5
B. 6
C. 7
D. 8

3. Subtract the double fact. 10 - 5 = ______

A. 5
B. 10
C. 15
D. 20

4. What is the number two written as a digit?

A. 2
B. 4
C. 6
D. 8

5. What is the missing fact in the fact family 6 + 1 = 7, 1 + 6 = 7, 7 - 1 = 6?

A. 7 - 7 = 0
B. 7 - 6 = 1
C. 7 - 2 = 5
D. 6 - 1 = 5

6. Counting forward, what number comes after 44?

A. 45
B. 46
C. 50
D. 55

GO ON

7. Fill in the blank. 1+ ______ = 16

A. 13
B. 14
C. 15
D. 16

8. How many ones are in 18?

A. 1
B. 2
C. 6
D. 8

9. What time is it?

A. 6 : 30
B. 8 : 30
C. 10 : 30
D. 12 : 30

10. Subtract. 9 - 1 = ______

A. 4
B. 6
C. 8
D. 10

11. Add. 1 + 3 = ______

A. 2
B. 4
C. 6
D. 8

12. What set of numbers completes the double fact 4 = ______ + ______?

A. 2, 2
B. 3, 3
C. 4, 4
D. 5, 5

GO ON ▶

13. Kay drew suns and moons on her paper and then graphed how many of each she drew. How many moons did Kay draw?

A. 7
B. 6
C. 4
D. 2

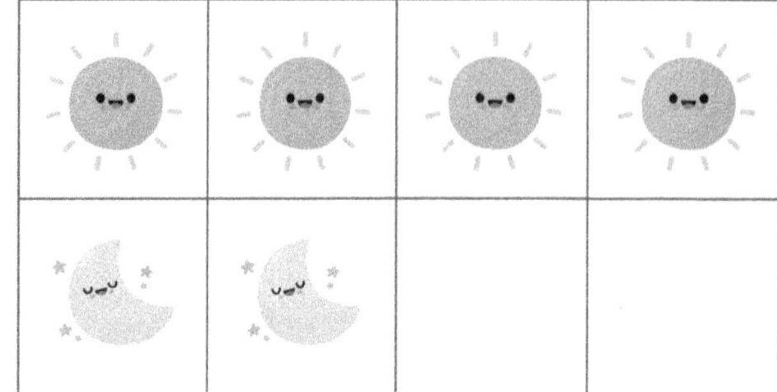

14. How do you make 10?

A. 7 + 2
B. 7 + 3
C. 7 + 4
D. 7 + 1

15. Fill in the blank. ______ + 7 = 8

A. 0
B. 1
C. 2
D. 3

16. What shape is this?

A. Sphere
B. Cone
C. Square
D. Cube

17. How many tens are in the diagram?

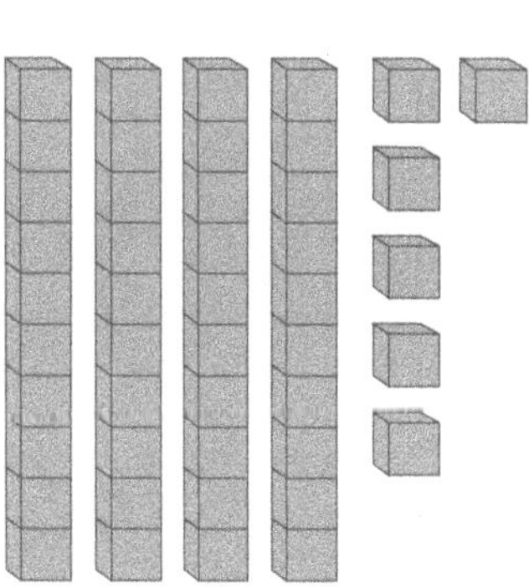

A. 6
B. 5
C. 4
D. 3

18. Which key is the longest? ____________________

A.

B.

C.

GO ON

19. What is the related subtraction fact 8 - 6 = 2?

A. 8 - 2 = 6
B. 8 - 8 = 0
C. 8 - 1 = 7
D. 2 - 2 = 0

20. Add the double fact. 4 + 4 = ______

A. 2
B. 4
C. 6
D. 8

21. What subtraction sentence is true?

A. 9 - 7 = 2
B. 9 - 6 = 2
C. 9 - 9 = 2
D. 9 - 5 = 5

22. What is 10 more than 17?

A. 7
B. 16
C. 27
D. 37

23. Add. 3 + 1 + 5 = ______

A. 10
B. 9
C. 8
D. 7

24. Jackie has 8 purple earrings and 2 green earrings. How many earrings does she have all together?

A. 8
B. 10
C. 12
D. 14

GO ON ▶

25. How many corners does a triangle have?

A. 1

B. 2

C. 3

D. 4

Session 2

GO ON

26. Fill in the blank. 7 - ______ = 0

A. 0
B. 3
C. 5
D. 7

27. What addition sentence is true?

A. 3 + 9 = 12
B. 3 + 9 = 11
C. 9 + 4 = 12
D. 9 + 2 = 12

28. How many ones are in this diagram?

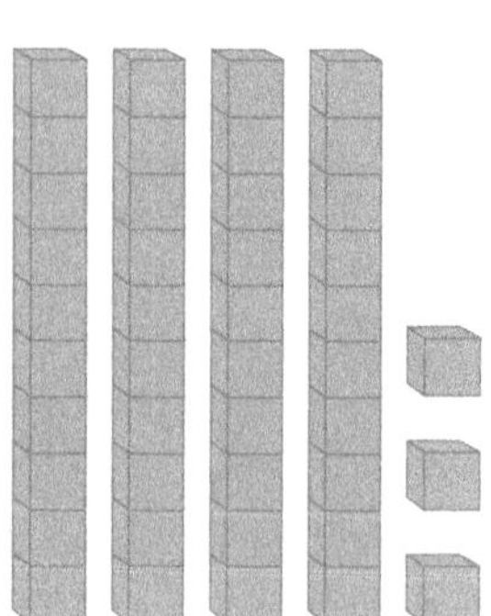

A. 3
B. 4
C. 5
D. 6

29. Add. 80 + 11 = ______

A. 81
B. 91
C. 71
D. 61

30. Add. 9 + 8 = ______

A. 15
B. 16
C. 17
D. 18

31. What number is 10 less than 23?

A. 13
B. 33
C. 43
D. 53

GO ON ▶

32. What is the related addition fact 2 + 7 = 9?

A. 7 + 1 = 8
B. 7 + 3 = 10
C. 7 + 2 = 9
D. 2 + 6 = 8

33. Subtract. 19 - 7 = ______

A. 12
B. 13
C. 14
D. 15

34. Ken saw 5 dogs at the dog park, 4 more joined them. How many were there now?

____

35. Fill in the missing number. If 5 + 2 = 7, then 7 - ______ = 5

36. How many tens are in 47? ______________________

37. What number comes after 67? ______________________

GO ON ▶

38. What is the missing number 19 - ______ = 16?

39. What is 7 written as a word? ______________________________

40. What symbol makes this sentence true 18 ______ 19?

ANSWER KEY

Question	Type	Key	Points	Standard	Cluster	Explanation
Session 1						
1	MC	C	1	CCSS.Math. Content.1.OA.A1	Operations and Algebraic Thinking	4 + 1 = 5
2	MC	B	1	CCSS.Math. Content.1.OA.C.6	Operations and Algebraic Thinking	2 + 5 = 7
3	MC	A	1	CCSS.Math. Content.1.OA.C.6	Operations and Algebraic Thinking	10 - 3 = 7
4	MC	A	1	CCSS.Math. Content.1.OA.C.6	Operations and Algebraic Thinking	9 - 5 = 4
5	MC	C	1	CCSS.Math. Content.1.OA.C.6	Operations and Algebraic Thinking	Write an addition sentence, 3 + 4 + 2 = 9
6	MC	D	1	CCSS.Math. Content.1.OA.C.6	Operations and Algebraic Thinking	3 + 3 = 6
7	MC	B	1	CCSS.Math. Content.1.OA.C.6	Operations and Algebraic Thinking	5 + 3 = 8
8	MC	C	1	CCSS.Math. Content.1.OA.C.6	Operations and Algebraic Thinking	7 - 0 = 7
9	MC	D	1	CCSS.Math. Content.1.OA.C.6	Operations and Algebraic Thinking	5 + 5 = 10
10	MC	C	1	CCSS.Math. Content.1.OA. D.8	Operations and Algebraic Thinking	8 = 4 + 4
11	MC	A	1	CCSS.Math. Content.1.OA. D.8	Operations and Algebraic Thinking	9 - 4 = 5
12	MC	B	1	CCSS.Math. Content.1.OA.C.6	Operations and Algebraic Thinking	2 + 3 = 5
13	MC	C	1	CCSS.Math. Content.1.OA.B.3	Operations and Algebraic Thinking	Fact families have two addition and two subtraction sentences that use the same 3 numbers. 8 - 6 = 2 is missing
14	MC	D	1	CCSS.Math. Content.1. NBT.A.1	Number and Operations in Base Ten	Five = 5
15	MC	A	1	CCSS.Math. Content.1.NBT.B.2	Number and Operations in Base Ten	There are two tens in 22. Meaning that you can make two groups of 10 with 2 ones remaining.
16	MC	B	1	CCSS.Math. Content.1. NBT.B.2	Number and Operations in Base Ten	There are 7 ones in 37. You can make three groups of ten with 7 ones left over.
17	MC	C	1	CCSS.Math. Content.1.NBT. C. 4	Number and Operations in Base Ten	40 + 11 = 51
18	MC	D	1	CCSS.Math. Content.1. NBT. C. 5	Number and Operations in Base Ten	30 - 20 = 10

Question	Type	Key	Points	Standard	Cluster	Explanation
19	MC	B	1	CCSS.Math. Content.1.NBT. C. 5	Number and Operations in Base Ten	10 + 7 = 17
20	MC	C	1	CCSS.Math. Content.1. NBT. C. 5	Number and Operations in Base Ten	34 - 10 = 24
21	MC	A	1	CCSS.Math. Content.1.NBT. C.4	Number and Operations in Base Ten	8 + 8 + 2 = 18 Add the doubles and then add 2
22	MC	D	1	CCSS.Math. Content.1.NBT.C.4	Number and Operations in Base Ten	Regrouping. 17 + 4 = 21
23	MC	D	1	CCSS.Math. Content.1.NBT. C. 4	Number and Operations in Base Ten	1 ten = 10 ones. You break down 1 ten into ones. Keeping the other two tens the same and add back the new ones. Leaving you with 2 tens and 14 ones.
24	MC	C	1	CCSS.Math. Content.1. MD. B.3	Measurement and Data	The clock reads 7 : 00. Small hand on the hour and large hand on the minute.
25	MC	C	1	CCSS.Math. Content.1. MD. C. 4	Measurement and Data	There are 5 oranges in the table.
Session 2						
26	MC	B	1	CCSS.Math. Content.1. MD. C.4	Measurement and Data	B has the longest eraser
27	MC	D	1	CCSS.Math. Content.1. MD. C. 4	Measurement and Data	14 students picked soccer
28	MC	C	1	CCSS.Math. Content.1. GA. 1	Geometry	C is a cube
29	MC	A	1	CCSS.Math. Content.1. GA.1	Geometry	A is a triangle and has 3 sides
30	MC	C	1	CCSS. Math. Content 1.OA.A1	Operations and Algebraic Thinking	There are 6 cars to start. If one more drives in there are a total of 7. 6 + 1 = 7
31	MC	A	1	CCSS. Math. Content. 1.OA. C.6	Operations and Algebraic Thinking	Add up all of the pencils. 3 + 5 + 2 = 10 pencils
32	MC	B	1	CCSS.Math. Content.1.OA.B.3	Operations and Algebraic Thinking	Fact families have two addition and two subtraction sentences that use the same 3 numbers. 5 - 3 = 2
33	MC	D	1	CCSS.Math. Content.1.MD. B. 3	Measurement and Data	The clock reads 4 : 30

GO ON ▶

Question	Type	Key	Points	Standard	Cluster	Explanation
34	SR	34 < 44	2	CCSS.Math. Content.1. NBT. 4	Number and Operations in Base Ten	44 is greater than 34
35	SR	1 ten and 16 ones	2	CCSS.Math. Content.1.NBT. 3	Number and Operations in Base Ten	1 ten =10 ones. You break down 1 ten into ones. Keeping the other ten the same and add back the new ones. Leaving you with 1 ten and 16 ones.
36	SR	$.10, 10 cents	2	CCSS.Math. Content.1. MD. 3b	Measurement and Data	5 pennies and one nickel = ten cents
37	SR	3	2	CCSS.Math. Content.1.OA. 4	Operations and Algebraic Thinking	9 - 6 = 3
38	SR	3 cookies	2	CCSS.Math. Content.1. OA. 1	Operations and Algebraic Thinking	8 - 5 = 3 cookies
39	SR	3 + 4 =7	2	CCSS.Math. Content.1. OA. 1	Operations and Algebraic Thinking	3 dogs + 4 dogs = 7 dogs
40	ER	$7	2	CCSS.Math. Content.1.OA 1	Operations and Algebraic Thinking	She gave $5 away for candy canes and has $2 left. 5 + 2 = 7

Question	Type	Key	Points	Standard	Cluster	Explanation
Session 1						
1	MC	C	1	CCSS.Math. Content.1.OA.C.1	Operations and Algebraic Thinking	4 + 4 = 8
2	MC	B	1	CCSS.Math. Content.1. OA. A. 1	Operations and Algebraic Thinking	9 - 9 = 0
3	MC	C	1	CCSS.Math. Content.1.NBT. A. 1	Number and Operations in Base Ten	Counting on from 72 the next number is 73
4	MC	D	1	CCSS.Math. Content.1. NBT.B.2b	Number and Operations in Base Ten	There are 6 ones in 16
5	MC	D	1	CCSS.Math. Content.1.NBT.A.1	Number and Operations in Base Ten	There are 8 stars
6	MC	A	1	CCSS.Math. Content.1. OA. 4	Operations and Algebraic Thinking	If 4 + 5 = 9, then 9 - 5 = 4. The other sentences are incorrect.
7	MC	D	1	CCSS. Math. Content. 1.NBT.A.1	Number and Operations in Base Ten	TEN = 10
8	MC	B	1	CCSS.Math. Content.1.NBT. C. 6	Number and Operations in Base Ten	90 - 10 = 80
9	MC	C	1	CCSS.Math. Content.1.NBT. C. 4	Number and Operations in Base Ten	5 + 2 + 5 = 12
10	MC	C	1	CCSS.Math. Content.1. OA. A. 1	Operations and Algebraic Thinking	3 + 2 = 5
11	MC	A	1	CCSS.Math. Content.1. MB. B. 3	Measurement and Data	5 : 00 matches the analog clock
12	MC	B	1	CCSS.Math. Content.1. OA.C.6	Operations and Algebraic Thinking	13 - 13 = 0
13	MC	C	1	CCSS.Math. Content.1.OA. A.1	Operations and Algebraic Thinking	14 - 7 = 7
14	MC	B	1	CCSS.Math. Content.1. OA.D.7	Operations and Algebraic Thinking	10 - 6 = 4
15	MC	A	1	CCSS.Math. Content.1.OA. C. 6	Operations and Algebraic Thinking	3 = 10 - 7
16	MC	A	1	CCSS.Math. Content.1. MD. 3	Measurement and Data	It is a nickel
17	MC	D	1	CCSS.Math. Content.1.OA.B.3	Operations and Algebraic Thinking	4 + 2 + 1 = 7
18	MC	D	1	CCSS.Math. Content.1.NBT.C.4	Number and Operations in Base Ten	20 + 40 = 60
19	MC	A	1	CCSS.Math. Content.1.OA.D.8	Operations and Algebraic Thinking	10 = 5 + 5
20	MC	B	1	CCSS.Math. Content.1. NBT. A. 1	Number and Operations in Base Ten	17 = seventeen

GO ON ▶

Question	Type	Key	Points	Standard	Cluster	Explanation
21	MC	D	1	CCSS.Math. Content.1.NBT. B. 2	Number and Operations in Base Ten	There are 4 groups of ten in 41
22	MC	A	1	CCSS.Math. Content.1.NBT. B. 2	Number and Operations in Base Ten	There is one, one in 41
23	MC	A	1	CCSS.Math. Content.1. OA.A.1	Operations and Algebraic Thinking	3 + 4 = 7
24	MC	A	1	CCSS.Math. Content.1.OA.D.7	Operations and Algebraic Thinking	4 + 7 = 11
25	MC	D	1	CCSS.Math. Content.1. MD.3	Measurement and Data	There are 3 pennies and 1 dime adding up to $.13
Session 2						
26	MC	C	1	CCSS.Math. Content.1. NBT. C. 4	Number and Operations in Base Ten	12 + 4 = 16
27	MC	C	1	CCSS.Math. Content.1. OA. D.8	Operations and Algebraic Thinking	10 - 6 = 4
28	MC	A	1	CCSS.Math. Content.1.OA.C.6	Operations and Algebraic Thinking	7 + 8 = 15
29	MC	B	1	CCSS.Math. Content.1. OA.B.4	Operations and Algebraic Thinking	If 3 + 6 = 9, the 9 - 6 = 3
30	MC	D	1	CCSS. Math. Content 1.OA.B.3	Operations and Algebraic Thinking	6 + 2 + 3 = 11
31	MC	C	1	CCSS. Math. Content. 1. OA.A2	Operations and Algebraic Thinking	3 + 2 = 5 = 10
32	MC	A	1	CCSS.Math. Content.1.OA.C.6	Operations and Algebraic Thinking	17 - 17 = 0
33	MC	B	1	CCSS.Math. Content.1.OA. D. 8	Operations and Algebraic Thinking	2 + 5 = 7
34	SR	=	2	CCSS.Math. Content.1. NBT. B. 3	Operations and Algebraic Thinking	=, 34 = 34
35	SR	Cube	2	CCSS.Math. Content.1.MD. 3	Measurement and Data	Cube
36	SR	4 : 00	2	CCSS.Math. Content.1. MD. B. 3	Measurement and Data	The clock reads 4 : 00
37	SR	3 + 1 = 4	2	CCSS.Math. Content.1. OA.B.1	Operations and Algebraic Thinking	Fact families have four sets of facts using 3 numbers. The missing set is 3 + 1 = 4

Question	Type	Key	Points	Standard	Cluster	Explanation
38	SR	36	2	CCSS.Math. Content.1. NBT.C.5	Number and Operations in Base Ten	26 + 10 = 36
39	SR	5	2	CCSS.Math. Content.1. OA. A.1	Operations and Algebraic Thinking	2 + 3 = 5, 2 hamsters + 3 guinea pigs = 5 pets
40	ER	17 - 9 = 8	2	CCSS.Math. Content.1.OA.A.1	Operations and Algebraic Thinking	Jack has 17 teddy bears then gives away 9. 17 - 9 = 8

GO ON ▶

Question	Type	Key	Points	Standard	Cluster	Explanation
Session 1						
1	MC	A	1	CCSS.Math. Content.1.NBT.A.1	Number and Operations in Base Ten	7 = seven
2	MC	A	1	CCSS.Math. Content.1.NBT.A.1	Number and Operations in Base Ten	There are 6 hearts
3	MC	D	1	CCSS.Math. Content.1.OA.C.6	Operations and Algebraic Thinking	5 + 5 = 10
4	MC	C	1	CCSS.Math. Content.1NBT.B.2b	Number and Operations in Base Ten	There is 1 ten in 12
5	MC	A	1	CCSS.Math. Content.1. OA.A. 1	Operations and Algebraic Thinking	4 + 2= 6
6	MC	B	1	CCSS.Math. Content.1.NBT.C.4	Number and Operations in Base Ten	30 + 50 = 80
7	MC	A	1	CCSS. Math. Content. 1.OA.C.6	Operations and Algebraic Thinking	12 - 0 = 12
8	MC	D	1	CCSS.Math. Content.1.NBT. C. 4	Number and Operations in Base Ten	6 + 1 + 6 = 13
9	MC	D	1	CCSS.Math. Content.1.NBT. C.5	Number and Operations in Base Ten	54 + 10 = 64
10	MC	A	1	CCSS.Math. Content.1. OA.B.3	Operations and Algebraic Thinking	3 + 3 + 1 = 7
11	MC	D	1	CCSS.Math. Content.1.NBT. B. 2	Number and Operations in Base Ten	There are 5 groups of ten in 56
12	MC	B	1	CCSS.Math. Content.1.NBT.B.2	Number and Operations in Base Ten	There are 6 ones in 56
13	MC	A	1	CCSS.Math. Content.1.NBT. A. 1	Number and Operations in Base Ten	14 = fourteen
14	MC	C	1	CCSS.Math. Content.1.OA.D.7	Operations and Algebraic Thinking	14 - 4 = 10
15	MC	B	1	CCSS.Math. Content.1.MD.3	Measurement and data	Dime
16	MC	A	1	CCSS.Math. Content.1.OA.A.1	Operations and Algebraic Thinking	13 - 5 = 8
17	MC	A	1	CCSS.Math. Content.1. OA. D.7	Operations and Algebraic Thinking	6 + 5 = 11
18	MC	B	1	CCSS.Math. Content.1.OA.D. 8	Operations and Algebraic Thinking	7 - 6 = 1
19	MC	D	1	CCSS.Math. Content.1. OA. C 6	Operations and Algebraic Thinking	5 = 9 - 4
20	MC	B	1	CCSS.Math. Content.1.NBT. A. 1	Number and Operations in Base Ten	Counting forward 50 is the next number
21	MC	B	1	CCSS.Math. Content.1.OA. C. 6	Operations and Algebraic Thinking	9 = 5 + 4

Question	Type	Key	Points	Standard	Cluster	Explanation
22	MC	B	1	CCSS.Math. Content.1.OA. A.1	Operations and Algebraic Thinking	17 - 8 = 9
23	MC	B	1	CCSS.Math. Content.1.NBT. B.3	Number and Operations in Base Ten	33 < 43
24	MC	C	1	CCSS.Math. Content.1. OA. D.8	Operations and Algebraic Thinking	8 = 4 + 4
25	MC	C	1	CCSS.Math. Content.1.MB.B.3	Measurement and Data	The clock reads 3 : 30
Session 2						
26	MC	B	1	CCSS.Math. Content.1.OA.C.6	Operations and Algebraic Thinking	11 + 9 = 20
27	MC	A	1	CCSS.Math. Content.1.OA. B.4	Operations and Algebraic Thinking	8 - 3 = 5, then 5 + 3 = 8
28	MC	D	1	CCSS.Math. Content.1OA. C. 6	Operations and Algebraic Thinking	2 + 3 + 4 = 9
29	MC	B	1	CCSS.Math.Content.1 OA.C.6	Operations and Algebraic Thinking	9 + 1 = 10
30	MC	A	1	CCSS. Math. Content 1.OA.C.6	Operations and Algebraic Thinking	18 - 9 = 9
31	MC	D	1	CCSS. Math. Content. 1.OA.D.8	Operations and Algebraic Thinking	2 + 7 = 9
32	MC	D	1	CCSS.Math. Content.1. NBT. C. 4	Number and Operations in Base Ten	11 + 6 = 17
33	MC	C	1	CCSS.Math. Content.1. OA. A 2	Operations and Algebraic Thinking	3 + 4 + 3 = 10
34	SR	Triangle	2	CCSS.Math. Content.1. MD. 3	Measurement and Data	Triangles have three corners
35	SR	$.15	2	CCSS.Math. Content.1.MD. 3	Measurement and Data	There is one dime and one nickel adding up to $.15
36	SR	7 : 00	2	CCSS.Math. Content.1.MB. B. 3	Measurement and Data	The clock reads 7 : 00
37	SR	60	2	CCSS.Math. Content.1. NBT.C.6	Number and Operations in Base Ten	80 - 20 = 60
38	SR	2 + 3 = 5	2	CCSS.Math. Content.1. OA.B.1	Operations and Algebraic Thinking	Fact families have four facts using 3 numbers. The missing fact is 2 + 3 = 5
39	SR	4	2	CCSS.Math. Content.1. OA. 4	Operations and Algebraic Thinking	10 + 4 = 14, therefore 14 - 4 = 10. 4 is the missing number.
40	ER	16 - 10 = 6	2	CCSS.Math. Content.1.OA.A.1	Operations and Algebraic Thinking	16 - 10 = 6, 16 rubber duckies start in the tub then subtract 10 that stay, 6 go to bed with Tommy.

GO ON ▶

Question	Type	Key	Points	Standard	Cluster	Explanation
Session 1						
1	MC	C	1	CCSS.Math. Content.1.NBT.B.2b	Number and Operations in Base Ten	There are 9 ones in 19
2	MC	A	1	CCSS.Math. Content.1.NBT.C.5	Number and Operations in Base Ten	41 + 10 = 51
3	MC	C	1	CCSS.Math. Content.1.NBT.A.1	Number and Operations in Base Ten	Nine = 9
4	MC	B	1	CCSS.Math. Content.1. OA.B1	Operations and Algebraic Thinking	1 + 3 + 5 = 9
5	MC	C	1	CCSS.Math. Content.1. OA.D. 7	Operations and Algebraic Thinking	9 - 3 = 6
6	MC	C	1	CCSS.Math. Content.1. NBT.C. 6	Number and Operations in Base Ten	40 - 10 = 30
7	MC	B	1	CCSS. Math. Content. 1. OA. A. 1	Operations and Algebraic Thinking	7 - 4 = 3
8	MC	C	1	CCSS.Math. Content.1.NBT.A.1	Number and Operations in Base Ten	12 = twelve
9	MC	D	1	CCSS.Math. Content.1.MB. B. 3	Measurement and Data	The clock reads 5 : 00
10	MC	A	1	CCSS.Math. Content.1.NBT.C.4	Number and Operations in Base Ten	2 + 2 + 7 = 11
11	MC	C	1	CCSS.Math. Content.1. NBT. C. 4	Number and Operations in Base Ten	10 + 40 = 50
12	MC	C	1	CCSS.Math. Content.1.NBT.A.1	Number and Operations in Base Ten	There are 9 x's
13	MC	A	1	CCSS.Math. Content.1.OA. C.6	Operations and Algebraic Thinking	7 + 7 = 14
14	MC	D	1	CCSS.Math. Content.1.OA. A.1	Operations and Algebraic Thinking	3 + 3 = 6
15	MC	C	1	CCSS.Math. Content.1.OA. C. 6	Operations and Algebraic Thinking	4 = 10 - 6
16	MC	D	1	CCSS.Math. Content.1.MB. B. 3	Measurement and Data	The analog clock reads 4 : 30
17	MC	C	1	CCSS.Math. Content.1.NBT.A.1	Number and Operations in Base Ten	36 is the next number in the counting sequence
18	MC	B	1	CCSS.Math. Content.1.OA.C.6	Operations and Algebraic Thinking	14 - 1 = 13
19	MC	B	1	CCSS.Math. Content.1.OA.D.8	Operations and Algebraic Thinking	3 + 3 = 6
20	MC	A	1	CCSS.Math. Content.1.OA.D.7	Operations and Algebraic Thinking	5 - 5 = 0
21	MC	B	1	CCSS.Math. Content.1.OA.D.8	Operations and Algebraic Thinking	1 + 9 = 10

Question	Type	Key	Points	Standard	Cluster	Explanation
22	MC	D	1	CCSS.Math. Content.1.OA.D.8	Operations and Algebraic Thinking	3 + 11 = 14
23	MC	A	1	CCSS.Math. Content.1.OA. C.6	Operations and Algebraic Thinking	5 + 2 = 7, then 7 - 5 = 2
24	MC	C	1	CCSS.Math. Content.1.OA. B. 3	Measurement and Data	Penny
25	MC	D	1	CCSS.Math. Content.1. OA. A.1	Operations and Algebraic Thinking	19 - 9 = 10
Session 2						
26	MC	C	1	CCSS.Math. Content.1. OA. C.6	Operations and Algebraic Thinking	5 = 3 + 2
27	MC	A	1	CCSS.Math. Content.1. OA. A. 2	Operations and Algebraic Thinking	3 + 4 + 1 = 8
28	MC	A	1	CCSS.Math. Content.1.NBT	Number and Operations in Base Ten	79 is the next number in the sequence
29	MC	B	1	CCSS.Math. Content.1. NBT. C. 4	Number and Operations in Base Ten	14 + 5 = 19
30	MC	C	1	CCSS. Math. Content 1. OA.C. 6	Operations and Algebraic Thinking	4 + 2 + 1 = 7
31	MC	D	1	CCSS. Math. Content. 1. NBT. C. 6	Number and Operations in Base Ten	20 - 10 = 10
32	MC	A	1	CCSS.Math. Content.1.NBT. B.2	Number and Operations in Base Ten	There are 2 groups of tens in 29
33	MC	D	1	CCSS.Math. Content.1.NBT. B.2	Number and Operations in Base Ten	There are 9 ones in 29
34	SR	$.26	2	CCSS.Math. Content.1.MD. 3	Measurement and Data	There is one quarter and one penny adding up to $.26
35	SR	7	2	CCSS.Math. Content.1.OA. 4	Operations and Algebraic Thinking	10 - 3 = 7
36	SR	5	2	CCSS.Math. Content.1.OA. D. 8	Operations and Algebraic Thinking	7 - 5 = 2
37	SR	4	2	CCSS.Math. Content.1. MD.3	Measurement and Data	The shape has 4 corners
38	SR	7 - 4 = 3	2	CCSS.Math. Content.1. OA.B.1	Operations and Algebraic Thinking	Fact families have four facts using 3 numbers. The missing fact in this group is 7 - 4 = 3
39	SR	17	2	CCSS.Math. Content.1. OA.A.1	Operations and Algebraic Thinking	13 + 4 = 17
40	ER	>	2	CCSS.Math. Content.1.NBT.B.3	Number and Operations in Base Ten	55 > 53

GO ON ▶

Question	Type	Key	Points	Standard	Cluster	Explanation
Session 1						
1	MC	B	1	CCSS.Math. Content.1. OA. C.6	Operations and Algebraic Thinking	4 + 4 = 8
2	MC	B	1	CCSS.Math. Content.1. OA. C. 6	Operations and Algebraic Thinking	13 - 4 = 9
3	MC	A	1	CCSS.Math. Content.1.NBT.A.1	Number and Operations in Base Ten	19 > 17
4	MC	D	1	CCSS.Math. Content.1.OA.B	Operations and Algebraic Thinking	10 - 2 = 8
5	MC	C	1	CCSS.Math. Content.1OA.D.8	Operations and Algebraic Thinking	14 = 7 + 7
6	MC	A	1	CCSS.Math. Content.1.OA.C.6	Operations and Algebraic Thinking	4 + 3 = 7
7	MC	C	1	CCSS. Math. Content. 1.OA.C.6	Operations and Algebraic Thinking	If 3 + 5 = 8, then 8 - 5 = 3
8	MC	C	1	CCSS.Math. Content.1.OA.C.6	Operations and Algebraic Thinking	4 + 9 = 13
9	MC	B	1	CCSS.Math. Content.1.OA.B	Operations and Algebraic Thinking	If 10 - 6 = 4, then 10 - 4 = 6
10	MC	D	1	CCSS.Math. Content.1.NBT.A.1	Number and Operations in Base Ten	Eight = 8
11	MC	C	1	CCSS.Math. Content.1.NBT.A.1	Number and Operations in Base Ten	The next number is 41
12	MC	B	1	CCSS.Math. Content.1.OA.C.6	Operations and Algebraic Thinking	2 + 2 + 4 = 8
13	MC	C	1	CCSS.Math. Content.1.OA.D.7	Operations and Algebraic Thinking	3 + 4 = 7
14	MC	A	1	CCSS.Math. Content.1.NBT.B.2	Number and Operations in Base Ten	There are 4 groups of ten in 46.
15	MC	A	1	CCSS.Math. Content.1.OA.B.4	Operations and Algebraic Thinking	7 + 2 = 9
16	MC	B	1	CCSS.Math. Content.1.NBT.C.5	Number and Operations in Base Ten	10 more than 57 is 67
17	MC	A	1	CCSS.Math. Content.1.OA.D.7	Operations and Algebraic Thinking	12 - 1 = 11
18	MC	C	1	CCSS.Math. Content.1.NBT.B.2	Number and Operations in Base Ten	There are 5 ones in 15.
19	MC	D	1	CCSS.Math. Content.1.NBT.C.4	Number and Operations in Base Ten	20 + 15 = 35
20	MC	C	1	CCSS.Math. Content.1.MD.B.3	Measurement and Data	It is 2 : 30
21	MC	C	1	CCSS.Math. Content.1.OA.B.4	Operations and Algebraic Thinking	8 - 8 = 0

Question	Type	Key	Points	Standard	Cluster	Explanation
22	MC	C	1	CCSS.Math. Content.1.MD.C.4	Measurement and Data	There are 3 cows
23	MC	D	1	CCSS.Math. Content.1.OA.C.6	Operations and Algebraic Thinking	8 - 3 = 5
24	MC	C	1	CCSS.Math. Content.1.OA.C.6	Operations and Algebraic Thinking	9 - 0 = 9
25	MC	B	1	CCSS.Math. Content.1.GA.1	Geometry	Triangles have 3 vertices
Session 2						
26	MC	C	1	CCSS.Math. Content.1. OA. C. 6	Operations and Algebraic Thinking	14 - 7 = 7
27	MC	C	1	CCSS.Math. Content.1.NBT.C.5	Number and Operations in Base Ten	54 is 10 less than 64
28	MC	D	1	CCSS.Math. Content.1.OA.A.1	Operations and Algebraic Thinking	3 + 6 = 9
29	MC	D	1	CCSS.Math. Content.1.GA.1	Geometry	The shape has 6 sides and is a hexagon
30	MC	A	1	CCSS. Math. Content 1. OA. C.6	Operations and Algebraic Thinking	10 + 2 = 12
31	MC	B	1	CCSS. Math. Content. 1.NBT.A.1	Number and Operations in Base Ten	There are 4 ten rods in the image.
32	MC	A	1	CCSS.Math. Content.1.MD.C.4	Measurement and Data	The first screw is the longest
33	MC	C	1	CCSS.Math. Content.1.OA.B	Operations and Algebraic Thinking	If 2 + 8 = 10, then 8 + 2 = 10
34	SR	50	2	CCSS.Math. Content.1.NBT.A.1	Number and Operations in Base Ten	50 is the next number in the counting sequence.
35	SR	10	2	CCSS.Math. Content.1.OA.A.1	Operations and Algebraic Thinking	15 - 5 = 10
36	SR	4	2	CCSS.Math. Content.1. OA.B.4	Operations and Algebraic Thinking	7 + 4 = 11
37	SR	five	2	CCSS.Math. Content.1. NBT.A.1	Number and Operations in Base Ten	5 = FIVE
38	SR	2	2	CCSS.Math. Content.1. NBT.C.4	Number and Operations in Base Ten	There are 2 groups of tens in the diagram that are blue.
39	SR	9	2	CCSS.Math. Content.1. OA.A.1	Operations and Algebraic Thinking	7 + 2 = 9
40	ER	5 - 1 = 4	2	CCSS.Math. Content.1.OA.B	Operations and Algebraic Thinking	Fact families have four fact groups using 3 numbers. The missing fact group is 5 - 1 = 4

GO ON ▶

Question	Type	Key	Points	Standard	Cluster	Explanation
Session 1						
1	MC	C	1	CCSS.Math. Content.1.OA.C.6	Operations and Algebraic Thinking	10 + 9 = 19
2	MC	A	1	CCSS.Math. Content.1.OA.C.6	Operations and Algebraic Thinking	If 3 + 6 = 9, then 9 - 3 = 6
3	MC	B	1	CCSS.Math. Content.1.OA.C.6	Operations and Algebraic Thinking	2 + 6 = 8
4	MC	D	1	CCSS.Math. Content.1.OA.B	Operations and Algebraic Thinking	5 + 5 = 10
5	MC	A	1	CCSS.Math. Content.1.OA.B	Operations and Algebraic Thinking	9 - 2 = 7 is the missing fact in the group of 4 related facts.
6	MC	D	1	CCSS.Math. Content.1.OA.B	Operations and Algebraic Thinking	If 3 + 4 = 7, then 4 + 3 = 7
7	MC	B	1	CCSS. Math. Content. 1.OA.C.6	Operations and Algebraic Thinking	7 - 2 = 5
8	MC	D	1	CCSS.Math. Content.1.OA.C.6	Operations and Algebraic Thinking	12 - 6 = 6
9	MC	B	1	CCSS.Math. Content.1.NBT.A.1	Number and Operations in Base Ten	12 < 14
10	MC	A	1	CCSS.Math. Content.1.OA.B	Operations and Algebraic Thinking	12 + 8 = 20
11	MC	C	1	CCSS.Math. Content.1. NBT. A. 1	Number and Operations in Base Ten	56 is the next number in the counting sequence.
12	MC	D	1	CCSS.Math. Content.1.NBT.B.2	Number and Operations in Base Ten	There are 9 ones in 19
13	MC	B	1	CCSS.Math. Content.1.OA.D.7	Operations and Algebraic Thinking	13 + 4 = 17
14	MC	B	1	CCSS.Math. Content.1.NBT.C.5	Number and Operations in Base Ten	21 is 10 less than 31
15	MC	C	1	CCSS.Math. Content.1.OA.A.1	Operations and Algebraic Thinking	12 - 4 = 8
16	MC	A	1	CCSS.Math. Content.1.NBT.C.4	Number and Operations in Base Ten	30 + 17 = 47
17	MC	B	1	CCSS.Math. Content.1.OA.B	Operations and Algebraic Thinking	6 - 4 = 2
18	MC	D	1	CCSS.Math. Content.1.MD.B.3	Measurement and Data	The clock reads 8 : 30
19	MC	B	1	CCSS.Math. Content.1.NBT.A.1	Number and Operations in Base Ten	Eleven = 11
20	MC	D	1	CCSS.Math. Content.1.OA.A.1	Operations and Algebraic Thinking	8 + 9 = 17
21	MC	A	1	CCSS.Math. Content.1.MD.C.4	Measurement and Data	17 students picked raspberries

Question	Type	Key	Points	Standard	Cluster	Explanation
22	MC	C	1	CCSS.Math. Content.1.GA.1	Geometry	Basketballs are spheres
23	MC	B	1	CCSS.Math. Content.1.OA. C. 6	Operations and Algebraic Thinking	1 + 1 = 2
24	MC	C	1	CCSS.Math.Content.1	Operations and Algebraic Thinking	17 - 2 = 15
25	MC	B	1	CCSS.Math. Content.1.OA.C.6	Operations and Algebraic Thinking	6 + 7 = 13
Session 2						
26	MC	D	1	CCSS.Math. Content.1.OA.C.6	Operations and Algebraic Thinking	10 - 4 = 6
27	MC	B	1	CCSS.Math. Content.1.NBT.C.5	Number and Operations in Base Ten	10 more than 60 is 70
28	MC	D	1	CCSS.Math. Content.1.OA.C.6	Operations and Algebraic Thinking	1 + 0 + 1 = 2
29	MC	C	1	CCSS.Math. Content.1.MD.C.4	Measurement and Data	Cantaloupe C is the longest
30	MC	A	1	CCSS. Math. Content 1.OA.B	Operations and Algebraic Thinking	If 7 - 2 = 5, then 7 - 5 = 2
31	MC	C	1	CCSS. Math. Content. 1.OA.C.6	Operations and Algebraic Thinking	5 + 1 = 6
32	MC	D	1	CCSS.Math. Content.1.OA.D.8	Operations and Algebraic Thinking	18 − 9 + 9
33	MC	C	1	CCSS.Math. Content.1.GA.1	Geometry	Hexagons have 6 vertices
34	SR	Six	2	CCSS.Math. Content.1.NBT.A.1	Number and Operations in Base Ten	6 = six
35	SR	12	2	CCSS.Math. Content.1.OA.B	Operations and Algebraic Thinking	12 - 12 = 0
36	SR	88	2	CCSS.Math. Content.1.NBT.A.1	Number and Operations in Base Ten	88 is the next number in the counting sequence.
37	SR	7	2	CCSS.Math. Content.1. NBT.B.2	Number and Operations in Base Ten	There are 7 tens in 73
38	SR	3	2	CCSS.Math. Content.1. NBT. C.4	Number and Operations in Base Ten	There are 3 ones in the diagram
39	SR	4	2	CCSS.Math. Content.1. NBT.C.4	Numbers and Operations in Base Ten	There are 4 groups of 10 in the diagram
40	ER	2	2	CCSS.Math. Content.1.OA.A.1	Operations and Algebraic Thinking	9 - 7 = 2

GO ON ▶

Question	Type	Key	Points	Standard	Cluster	Explanation
Session 1						
1	MC	B	1	CCSS.Math. Content.1.OA.C.6	Operations and Algebraic Thinking	6 - 3 = 3
2	MC	B	1	CCSS.Math. Content.1.NBT.A.1	Number and Operations in Base Ten	14 < 15
3	MC	B	1	CCSS.Math. Content.1.OA.D.8	Operations and Algebraic Thinking	10 = 5 + 5
4	MC	B	1	CCSS.Math. Content.1.B.2b	Number and Operations in Base Ten	Nine = 9
5	MC	C	1	CCSS.Math. Content.1.OA.D.7	Operations and Algebraic Thinking	18 + 2 = 20
6	MC	D	1	CCSS.Math. Content.1.OA.C.6	Operations and Algebraic Thinking	16 - 10 = 6
7	MC	B	1	CCSS. Math. Content. 1.OA.B	Operations and Algebraic Thinking	17 - 7 = 10
8	MC	B	1	CCSS.Math. Content.1.OA.C.6	Operations and Algebraic Thinking	1 + 4 = 5
9	MC	C	1	CCSS.Math. Content.1.OA.C.6	Operations and Algebraic Thinking	4 + 1 + 3 = 8
10	MC	C	1	CCSS.Math. Content.1.OA. C. 6	Operations and Algebraic Thinking	8 + 8 = 16
11	MC	D	1	CCSS.Math. Content.1.OA.B	Operations and Algebraic Thinking	9 + 1 = 10 I the missing fact.
12	MC	B	1	CCSS.Math. Content.1.OA. C.6	Operations and Algebraic Thinking	10 - 2 = 8
13	MC	A	1	CCSS.Math. Content.1.OA.A.1	Operations and Algebraic Thinking	7 - 1 = 6
14	MC	D	1	CCSS.Math. Content.1. NBT.A.1	Number and Operations in Base Ten	100 is the next number in the counting sequence.
15	MC	D	1	CCSS.Math. Content.1.NBT.B.2	Number and Operations in Base Ten	There are 6 ones in 61.
16	MC	C	1	CCSS.Math. Content.1.NBT.C.4	Number and Operations in Base Ten	40 + 21 = 61
17	MC	A	1	CCSS.Math. Content.1.MD.B.3	Measurement and Data	The clock reads 1 : 30
18	MC	B	1	CCSS.Math. Content.1.OA.B	Operations and Algebraic Thinking	8 - 1 = 7
19	MC	B	1	CCSS.Math. Content.1.OA.A.1	Operations and Algebraic Thinking	4 + 7 = 11
20	MC	D	1	CCSS.Math. Content.1.OA.C.6	Operations and Algebraic Thinking	9 - 6 = 3
21	MC	D	1	CCSS.Math. Content.1.MD.C.4	Measurement and Data	Joe ate 5 hot dogs

Question	Type	Key	Points	Standard	Cluster	Explanation
22	MC	D	1	CCSS.Math. Content.1.OA.C.6	Operations and Algebraic Thinking	17 - 0 = 17
23	MC	B	1	CCSS.Math. Content.1. OA.B	Operations and Algebraic Thinking	If 5 + 1 = 6, then 1 + 5 = 6
24	MC	C	1	CCSS.Math. Content.1.NBT.C.5	Number and Operations in Base Ten	63 is 10 less than 73
25	MC	D	1	CCSS.Math. Content.1.OA.B	Operations and Algebraic Thinking	6 + 0 = 6
Session 2						
26	MC	D	1	CCSS.Math. Content.1.OA.C.6	Operations and Algebraic Thinking	9 + 3 = 12
27	MC	C	1	CCSS.Math. Content.1.GA.1	Geometry	Triangles have 3 edges
28	MC	A	1	CCSS.Math. Content.1.NBT.C.4	Number and Operations in Base Ten	There are 3 tens in the diagram.
29	MC	D	1	CCSS.Math. Content.1.NBT.C.4	Number and Operations in Base Ten	There are 7 ones in the diagram
30	MC	A	1	CCSS. Math. Content 1.GA.1	Geometry	Spheres have 0 edges
31	MC	B	1	CCSS. Math. Content. 1.MD.C.4	Measurement and Data	There were 9 votes for cats
32	MC	D	1	CCSS.Math. Content.1. NBT.C.5	Number and Operations in Base Ten	94 is ten more than 84
33	MC	A	1	CCSS.Math. Content.1.OA. C.6	Operations and Algebraic Thinking	20 - 10 = 10
34	SR	7	2	CCSS.Math. Content.1.OA.C.6	Operations and Algebraic Thinking	If 10 - 7 = 3, then 3 + 7 = 10
35	SR	30	2	CCSS.Math. Content.1.NBT.A.1	Number and Operations in Base Ten	30 is the next number in the counting sequence
36	SR	2	2	CCSS.Math. Content.1.OA.B	Operations and Algebraic Thinking	12 + 2 = 14
37	SR	ten	2	CCSS.Math. Content.1. NBT.A.1	Number and Operations in Base Ten	10 = ten
38	SR	6 - 4 = 2	2	CCSS.Math. Content.1. OA.B	Operations and Algebraic Thinking	If 6 - 2 = 4, then 6 - 4 = 2
39	SR	3 ones or 13 ones	2	CCSS.Math. Content.1. NBT.B.2	Number and Operations in Base Ten	There are 3 ones or 13 ones
40	ER	5	2	CCSS.Math. Content.1.OA.A.1	Operations and Algebraic Thinking	9 - 4 = 5

GO ON

Question	Type	Key	Points	Standard	Cluster	Explanation
Session 1						
1	MC	D	1	CCSS.Math. Content.1.OA.C.6	Operations and Algebraic Thinking	8 - 4 = 4
2	MC	C	1	CCSS.Math. Content.1.OA.A.1	Operations and Algebraic Thinking	12 - 5 = 7
3	MC	A	1	CCSS.Math. Content.1. OA. C.6	Operations and Algebraic Thinking	10 - 5 = 5
4	MC	A	1	CCSS.Math. Content.1.NBT.A.1	Number and Operations in Base Ten	Two = 2
5	MC	B	1	CCSS.Math. Content.1.OA.B	Operations and Algebraic Thinking	7 - 6 = 1 is the missing fact
6	MC	A	1	CCSS.Math. Content.1.NBT.A.1	Number and Operations in Base Ten	45 is the next number in the counting sequence.
7	MC	C	1	CCSS. Math. Content. 1.OA.B	Operations and Algebraic Thinking	1 + 15 = 16
8	MC	D	1	CCSS.Math. Content.1.NBT.B.2	Number and Operations in Base Ten	There are 8 ones in 18
9	MC	C	1	CCSS.Math. Content.1.MD.B.3	Measurement and Data	The clock reads 10 : 30
10	MC	C	1	CCSS.Math. Content.1.OA.C.6	Operations and Algebraic Thinking	9 - 1 = 8
11	MC	B	1	CCSS.Math. Content.1.OA.C.6	Operations and Algebraic Thinking	1 + 3 = 4
12	MC	A	1	CCSS.Math. Content.1.OA.D.8	Operations and Algebraic Thinking	4 = 2 + 2
13	MC	D	1	CCSS.Math. Content.1.MD.C.4	Measurement and Data	Kay drew 2 moons
14	MC	B	1	CCSS.Math. Content.1.OA.C.6	Operations and Algebraic Thinking	7 + 3 = 10
15	MC	B	1	CCSS.Math. Content.1.OA.B	Operations and Algebraic Thinking	1 + 7 = 8
16	MC	D	1	CCSS.Math. Content.1.GA.1	Geometry	Cubes are 3D and have 6 faces
17	MC	C	1	CCSS.Math. Content.1.NBT.C.4	Number and Operations in Base Ten	There are 4 tens in the diagram
18	MC	B	1	CCSS.Math. Content.1.MD.C.4	Measurement and Data	Key B is the longest
19	MC	A	1	CCSS.Math. Content.1.OA.B	Operations and Algebraic Thinking	If 8 - 6 = 2?
20	MC	D	1	CCSS.Math. Content.1. OA. C.6	Operations and Algebraic Thinking	4 + 4 = 8
21	MC	A	1	CCSS.Math. Content.1.OA.D.7	Operations and Algebraic Thinking	9 - 7 = 2

Question	Type	Key	Points	Standard	Cluster	Explanation
22	MC	C	1	CCSS.Math. Content.1.NBT.C.5	Number and Operations in Base Ten	27 is 10 more than 17
23	MC	B	1	CCSS.Math. Content.1.OA.C.6	Operations and Algebraic Thinking	3 + 1 + 5 = 9
24	MC	B	1	CCSS.Math. Content.1.OA.A.1	Operations and Algebraic Thinking	8 + 2 = 10
25	MC	C	1	CCSS.Math. Content.1.GA.1	Geometry	Triangles have 3 corners
Session 2						
26	MC	D	1	CCSS.Math. Content.1.OA.B	Operations and Algebraic Thinking	7 - 7 = 0
27	MC	A	1	CCSS.Math. Content.1.OA.D.7	Operations and Algebraic Thinking	3 + 9 = 12
28	MC	A	1	CCSS.Math. Content.1.NBT.C.4	Number and Operations in Base Ten	There are 3 ones in the diagram
29	MC	B	1	CCSS.Math. Content.1.NBT.C.4	Number and Operations in Base Ten	80 + 11 = 91
30	MC	C	1	CCSS. Math. Content 1.OA.C.6	Operations and Algebraic Thinking	9 + 8 = 17
31	MC	A	1	CCSS. Math. Content. 1.NBT.C.5	Number and Operations in Base Ten	13 is 10 less than 23
32	MC	C	1	CCSS.Math. Content.1.	Operations and Algebraic Thinking	If 2 + 7 = 9, then 7+ 2 = 9
33	MC	A	1	CCSS.Math. Content.1.OA.C.6	Operations and Algebraic Thinking	19 - 7 = 12
34	SR	9	2	CCSS.Math. Content.1.OA.A.1	Operations and Algebraic Thinking	5 + 4 = 9
35	SR	2	2	CCSS.Math. Content.1.OA.C.6	Operations and Algebraic Thinking	If 5 + 2 = 7, then 7 - 2 = 5
36	SR	4	2	CCSS.Math. Content.1.NBT.C.4	Number and Operations in Base Ten	There are 4 tens in 47
37	SR	68	2	CCSS.Math. Content.1. NBT.A.1	Number and Operations in Base Ten	68 is the next number in the counting sequence.
38	SR	3	2	CCSS.Math. Content.1. OA.B	Operations and Algebraic Thinking	19 - 3 = 16
39	SR	Seven	2	CCSS.Math. Content.1.NBT.A.1	Number and Operations in Base Ten	7 = seven
40	ER	<	2	CCSS.Math. Content.1.NBT.A.1	Number and Operations in Base Ten	18 < 19

Made in USA - North Chelmsford, MA
1121589_9781951048259
06.09.2020 1122